AUSTRIA SWITZERLAND

Welcome to Austria and Switzerland

John and Shirley Harrison

Collins
Glasgow and London

Cover Photographs

Van Phillips

(Austria) top left: Salzburg
top rt: Kärntnerstrasse, Wien
btm left: Tirol
btm rt: Heiligenblut
(Switzerland) left: Matterhorn
rt: Château de Chillon
btm: Alpenhorns

Photographs

Van Phillips

Regional Maps

Mike Shand, Kažia L. Kram, Iain Gerard

Town Plans

M. & R. Piggott

Illustrations

pp. 6–7 Peter Joyce
p. 19 Barry Rowe

First published 1983
Copyright © text: John and Shirley Harrison 1983
Copyright © maps: Wm. Collins Sons & Co. Ltd.
Published by William Collins Sons and Company Limited
Printed in Great Britain

ISBN 0 00 447322 1

HOW TO USE THIS BOOK

The contents of this book shows how the countries are divided up into tourist regions. The book is in two main sections; general information and gazetteer. The latter is arranged in the tourist regions with an introduction and a regional map (detail below left). There are also plans of main cities (detail below right). All main entries listed in the gazetteer are shown on the regional maps. Places to visit and leisure facilities available in each region and city are indicated by symbols. Main roads, railways and airports are shown on the maps.

 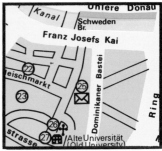

Regional Maps

- ◉ Museum/gallery
- ✝ Religious building
- ✈ Main airport
- ✈ Other airport
- ⛫ Castle/fortress
- ▲ Climbing
- ⌂ Interesting building
- ⚓ Boating/sailing
- ❀ Gardens
- ✺ Amusement park
- ⛷ Skiing/winter sports

- 🐘 Zoo
- m Ancient monument
- 🌳 Park
- ♨ Spa
- ⛏ Caves/mines
- 🚶 Walking/hiking
- 🐦 Bird watching

Town Plans

- ◉ Museum/gallery
- ✝ Religious building
- ⌂ Interesting building
- 🎭 Theatre
- 📚 Library
- ⛨ Town hall
- ✉ Post office
- ⛫ Castle/fortress
- ℹ Information
- POL Police
- 🌳 Park
- ❀ Garden
- ● Railway station
- 🚌 Bus station
- Ⓟ Car park

metres	feet
3000	9843
2000	6562
1000	3281
500	1640
200	656
0	0

AUSTRIA
1 : 1 500 000
0   10   20   30   40 kms
0   5   10   15   20   25 miles

SWITZERLAND
1 : 1 000 000
0   10   20   30 kms
0   5   10   15   20 miles

```
═══════  motorway
= = = =  motorway under construction
─────    other roads
─────    railway
```

Every effort has been made to give you an up-to-date text but changes are constantly occurring and we will be grateful for any information about changes you may notice while travelling.

CONTENTS

History	6
The Roof of Europe	8
The Arts	10
Paperwork	12
Customs	13
Currency	14
How to Get There	14
Internal Travel	16
If You Are Motoring	18
Accommodation	20
Food and Drink	22
Enjoy Yourself	24
Entertainment	26
What You Need to Know	28
Useful Addresses	31
Language	32
1 Wien	34
2 Eastern Austria	46
3 Southern Austria	52
4 Northern Austria	61
5 Western Austria	70
6 Southern Switzerland	80
7 Heartland Switzerland	92
8 Northern Switzerland	103
9 Western Switzerland	114
Index	124

Town Plans
1 Wien	36–7, 42–3
2 Bern	108–9
3 Genève	120–21

Regions
Provinces and Cantons

Switzerland
6 Graubünden 11
  Ticino 12
  Valais 13
7 Heartland Switzerland
  Uri 14
  Obwalden/Nidwalden 15
  Schwyz 16
  Glarus 17
  Zug 18
  Luzern 19
  Bern 20
8 Northern Switzerland
  St Gallen 21
  Appenzell 22
  Thurgau 23
  Zürich 24
  Schaffhausen 25
  Aargau 26
  Basel 27
  Solothurn 28
9 Western Switzerland
  Jura 29
  Neuchâtel 30
  Vaud 31
  Fribourg 32
  Genève 33

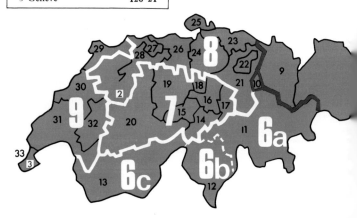

Austria
1 **Wien**
2 **Eastern Austria**
   Niederösterreich 1
   Burgenland 2
3 **Southern Austria**
   Steiermark 3
   Kärnten 4
   Osttirol 5
4 **Northern Austria**
   Oberösterreich 6
   Salzburg 7
5 **Western Austria**
   Tirol 8
   Vorarlberg 9
   Liechtenstein 10

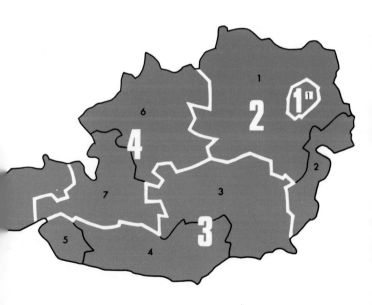

HISTORY

The Roman empire of Caesar Augustus haunted its barbarian destroyers for 400 years until Christmas Day, AD 800, when Charlemagne was crowned Karl I and founded the Holy Roman Empire. France, Germany and Italy were parts of that Empire and so were the Roman provinces of Helvetia and Rhaetia (modern Switzerland) and Noricum (modern Austria). Since then Switzerland has had fairly little history – at least none whose marks the casual visitor will observe, and the only name famous from Swiss history is that of the mythical William Tell. But the names and memorials of the makers of Austrian history are inescapable as you travel around; here is a breakdown of some.

Hohenstaufen The last real Emperor of the Holy Roman Empire was Frederick II, greatest of the Hohenstaufen dynasty, King of Germany, Sicily and Jerusalem. Around 1230 he granted a privilege to the men of the mountain around Andermatt, which started a movement that was to lead to the formation of Austria and Switzerland. He wanted a road across the Alps to unite Germany and Italy. The men of Uri provided it when they built a bridge across the Devil's Gorge near Göschenen and so made the St Gotthard Pass viable. (The first has long since vanished but from today's bridge you can still see the remains of the fourth and the ruins of the third.) In return Frederick granted them the right to have only the Emperor as their feudal overlord without intercession of duke, count, abbot or bishop. Since emperors are busy people this right meant, in effect, independence. The nominal overlords of Uri were the Counts of Habsburg and what they thought of this independence is not recorded.

Habsburgs Habichtsburg, the castle of the hawk, stands in ruins on a hill just south of Brugg in northern Switzerland. This was the family seat of the Habsburg family, whose young count Rudolf was godson of Frederick II. When Frederick died in 1250 there was no effective emperor for over 20 years, but in 1273 Rudolf was elected. The men of Uri realized that the rights granted by one emperor could be rescinded by another and they made a pact with their neighbours of Schwyz and Nidwalden to resist any attempt by Rudolf to resume his lost rights. But the new emperor had bigger fish to catch. East of his Swiss lands – today's Aargau and St Gallen – lay the eastern march of the Empire, today's province of Niederösterreich. This had been annexed by the King of Bohemia when the rightful rulers, the Babenbergs, died out (the Babenberg colours, red and white, were chosen as the flag of today's Austrian Republic). Rudolf put paid to the usurper, and in 1282 invested his sons as Dukes of Austria and Steiermark. Thus the Habsburgs started as they were to continue for over 600 years – personal glory in the name of the family.

Rudolf of Habsburg died in 1291 and his son claimed the title Emperor. Expecting the worst, the men of Uri renewed their alliance with Schwyz and Nidwalden and committed the terms in writing. By this document, still to be seen in the state archives in Schwyz, the allies agreed to obey their rightful lord but to accept no foreigner as 'judge'. To the Swiss, this is a hallowed charter of the foundation of the Swiss confederation, revered even more than the English Magna Carta; it is really the first declaration of Swiss neutrality.

The three forest cantons of Uri, Schwyz and Nidwalden were promptly joined by Luzern and Zurich but the following year the Habsburgs brought an army, smashed the little league at Winterthur and only Uri and its original partners stayed in revolt. The Habsburgs returned to subjugate all the family's old territories but this time the Austrians met the men of Schwyz armed with their secret weapon, the halberd, and in 1315 at Morgarten the Swiss achieved the first in an astonishing run of successes. From then on, more and more communities allied themselves to the original three and the Swiss confederation started to develop its modern shape, while Habsburg power west of the Rhine crumbled.

The Habsburgs collected other territories – Tirol, Kärnten, Vorarlberg – over the next 50 years, putting together the block of land that is modern Austria, but all were acquired by inheritance or purchase rather than force. Then they

were ready to try to regain their old lands in northern Switzerland; at Sempach in 1386 and at Näfels in 1388 huge Austrian armies were defeated or annihilated. You will see more reminders of these events west of the Rhine than east. The Aargau, where Castle Habsburg still stands, became subject to the confederation in the early 1400s.

Frederick III (1415–93) made the job of Emperor (of the Holy Roman Empire) a hereditary perk of the Habsburg ruler of Austria and this saddled his successors with the costly business of maintenance of status and prestige.

Maximilian I (died 1519) is recalled as the most glorious of the medieval emperors; he acquired Burgundy and the Netherlands by inheritance and lost Vienna by war; he also lost Switzerland for good, after defeat in the Swabian war, which ended with a treaty guaranteeing Swiss independence. (The Swabian war was part of a Swiss programme of empire building, which ended abruptly with their defeat at Marignono in 1515, and ever since then the Swiss have kept out of international politics.)

Ferdinand I inherited Hungary and Bohemia in the early 16th century, after driving the Turks away from Vienna; and the Danube lands of the Habsburgs stayed roughly the same up to the collapse of the monarchy in the 20th century. Under Ferdinand, Austria became almost wholly Protestant as the Reformation got under way, whereas in Switzerland the main cities (Geneva, under Calvin, Zurich, Basel and Bern) turned Protestant while the rural cantons stayed Catholic. The Swiss had little civil wars over religion but were too sensible not to see the benefits of compromise.

Rudolf II embraced the Counter-Reformation and the Habsburg lands were largely converted back to the Catholic faith, but at the cost of precipitating the Thirty Years' War (1618–48) which tore Europe apart. The Swiss, Catholic and Protestant, watched in peace and amazement and resolved on the religious tolerance which is so evident in Swiss life today. At the end of the war Switzerland was recognized as being outside the Holy Roman Empire and as permanently neutral.

Leopold I was Emperor when the Turks besieged Vienna in 1683; he called the forces of Christendom to his aid and Europe was relieved of the Turkish threat. That was the last time the idea of the Holy Roman Empire as the embodiment of a united Europe was alive, from then on the Habsburg Emperors were just the rulers of any territory they could hold down.

Maria Theresia became Empress in 1740 and Austrians look back on her reign as a Golden Age. The 18th-century air of most of Austria's old towns stems from her time as do the lighter aspects of the heavy wealth of Imperial architecture, such as the palace of Schönbrunn (west of Vienna).

Under Franz II the failures of the army against Napoleon forced the Emperor to abolish the Holy Roman Empire of the German Nation and entitle himself Franz I, Emperor of Austria. Napoleon reshaped other lands too – he broke up the Swiss Old Confederation, replaced it with a short-lived unitary Helvetic Republic and then built a new democratic confederation of 19 cantons – almost the same as today.

Franz-Josef became Emperor in 1848, when the loyal Viennese manned the barricades and cheered his uncle into abdication. ('Are they allowed to do that?' asked the departing Emperor when revolutionaries burst into the palace and killed the guards.)

The Swiss had had their revolution the year before, when the Catholic cantons of Inner Switzerland tried to secede. A short civil war followed in which 125 men were killed and the Swiss adopted a federal constitution which remains substantially the same today. From then on, a low profile became the Swiss ideal and all their energies have gone into precision and reliability in place of outward show.

Franz-Josef reigned for 68 years and there are still people who remember him affectionately as Der alte Herr – the old gentleman. Most of the grandiose buildings of Vienna and other large cities date from his reign, when the authorities tried to maintain the fiction that the Empire could hold together. No expense was spared to maintain the glitter and glamour of the monarchy, for prestige had always been the Habsburg ideal, but most people were aware that the days of the Empire were ending. Italians wanted to join the new Italy; Czechs, Poles, South Slavs demanded equality with Hungarians; Hungarians demanded equality with the Austrians and the Austrians turned to socialism or Germany. Unsuccessful wars led to loss of territory in Italy and the west and attempts to gain new Balkan territories in compensation led to the First World War.

The Habsburg monarchy was broken by defeat in that war and all the different nationalities of the Empire, except the German speakers, became independent or joined new independent countries; these are called the successor states in Austrian terminology.

The German-speaking provinces were left and these – the 'land nobody wanted' – came together in 1918 as the Austrian Republic. The rump of the Empire had no independent structure, no trade, no money, almost no food and the new Austrians set about building on the debts of the past while retaining some of the past graces. But above all they had no identity. Union with Germany, forbidden by the Allies, seemed the way for the Austrians to belong somewhere. Hitler marched in, in 1938, and nearly all Austria welcomed him. The welcome did not last and the Nazis treated Austrians as second-class Germans. You can see Austrian ambivalence in the war memorials honouring equally the many who died for Führer and Vaterland and the few who died as victims of the Gestapo.

Austria was 'liberated' by the Russians in 1945 and for the next ten years was occupied by the Allies. When the occupiers left, the Austrians started building the country anew.

It was committed by treaty to permanent neutrality, like Switzerland. Like Switzerland it has had to rely on making a living by choosing the right products made to a high quality with precision and reliability (reliability is not the characteristic normally associated with the old image of the gay, worldly Austrian), and like Switzerland it is a democratic republic ruled only by the mountains. As Switzerland and Austria evolve it will be interesting to see if they come to share the same character.

THE ROOF OF EUROPE

The Alps form only a minor part of France and Italy, Yugoslavia and Germany, but they dominate the two smaller countries, Austria and Switzerland, which these greater lands surround. The mountains bring most people here to enjoy the peaks, lakes and forests, mountain sports, clean air and local hospitality. There are other aspects of each country, but it is the mountains which determine their character and make them almost one. Superficial features which arise from this setting are common to both: the 'Swiss chalet' style of building – a rectangle with double-pitched roof and decorated balcony across a gabled end; the costume, called by outsiders 'Tirolean', of shorts with bib front for the men, colourful skirt and pinafore for the women; and the exceptionally high standard of cleanliness which is both an experience and a lesson. A more important characteristic is *Gemütlichkeit*, a warm, small-scale friendliness and contentment. Somehow this attribute has become associated with lowland Vienna, but it is a necessity of existence in mountain conditions, and you find it in most alpine inns, from where it has spread to the towns.

In addition to the mountains, there are of course the old towns, the fun towns, the musical towns, rolling pastoral landscapes, flat landscapes, industry and culture. But never out of sight are the picture-postcard peaks. Highest in Switzerland is Monte Rosa, 4634m/15,060ft, and in Austria the Grossglockner, 3797m/12,340ft.

Switzerland and Austria have roughly the same populations – 6.3 million in Switzerland, 7.5 million in Austria. But the densities are very different – Switzerland's 41,295sqkm/15,944sqmi (about half the area of Ireland) has 153 inhabitants per sqkm (Britain has 229), while Austria's 84,800sqkm/32,375sqmi has only 90 inhabitants per sqkm – less than France.

About 65% of the area of Switzerland, and 75% of Austria, is mountain, and all the mountain areas and much of the lower land has snow from December to April, and in places the snow lies from October to June. In the spring thaw alpine flowers appear and this is a favourite time for holidaymakers. Throughout the summer there is bright sunshine in the mountains interspersed with occasional rain or mist especially in the northern Alps; in the afternoons as the warmed air rises, clouds form over the mountain peaks, a sign of good weather. In the autumn when the air is filled with the scent of drying hay there tend to be longer spells of rain, and November can be a wet miserable month. But within this general picture there are huge variations; the glaciers above the line of perpetual snow are cold in summer and arctic in winter (1995sqkm/770sqmi of glaciers in Switzerland, compared with 1347sqkm/520sqmi of lakes); the cold of the Jura winter has to be felt to be believed even though the thermometer may seem to contradict; there is Mediterranean warmth in southern Switzerland and a continental swing from heavy summer heat to winter cold in eastern Austria.

In the main, tourists go to Austria and Switzerland for the scenery, the air and the alpine sports. Yet beneath this picture-postcard façade there is great variety and richness of interest. There are the monasteries and Baroque towns of Austria, many with their medieval core intact, as well as a vast store of cultural and artistic wealth. Switzerland has elegant resorts, colour and sense of quiet calm even in the busiest places.

Both countries are Federal Republics. Switzerland is made up of 26 cantons, which originally came together as a confederation, an alliance of free states. It has been a federation since 1848 and the individual cantons have more independence than the separate states of the USA; its official title is still the 'Swiss Confederation'. The official (Latin) name of the Swiss Confederation is *Confoederatio Helvetica* which explains the CH seen on Swiss cars and in post codes. Swiss citizenship derives from citizenship of the canton. About one sixth of the resident population of Switzerland are not citizens. Austria was formed, in 1919, out of nine German-speaking provinces, part of the disintegrated Habsburg empire. The provinces have considerable autonomy, but central government in Austria is considerably stronger than that of Switzerland.

The Austrians are about 90% Catholic, 6% Protestant, the province of Salzburg with its mining tradition is strongly Protestant despite all the efforts of its prince-archbishops. A small majority of Swiss citizens is Protestant, but Catholics are a small majority of all Swiss residents. Most cantons adhere officially to one denomination or the other, while giving complete freedom and equality to the minority denomination, while a few cantons are *paritätisch*, that is they accord total equality to both denominations.

A very interesting aspect of Swiss neutrality is the army. Every man has to perform military service from age 18 to 20 and then remains in the reserve for 30 years, doing a two-week period of military training every year. In the mountains you will often see the Swiss army on training manoeuvres. Every reservist keeps his uniform, rifle and ammunition at home, ready for instant recall to duty if the country is threatened. There has never been a recall since the system was instituted, though it came close, in 1940, when General Guisan assembled every officer in the army at the sacred field of Rutli (where the confederation was founded) to exhibit defiance against Hitler. Surely, no other country would leave its citizens in the possession of arms – yet there has been no instance of the rifles being used for private purposes.

Both countries are part of the western economic system. Switzerland is outwardly pure capitalist but its free enterprise is tempered by a discreet distribution of wealth to buy the cooperation of the trades unions, while Austria is an evolving, empirical blend of social welfare and individual self-help. But they remain outside alliances such as NATO and the European Economic Community, as they are politically neutral. Switzerland has kept out of major conflicts for over 400 years and has been internationally recognized as neutral since 1815; Austria welcomed its own neutrality when the Allies withdrew in 1955.

To assist tourists, and encourage foreign visitors to spend some of their money in Austria or Switzerland, each country has a central tourist office called in this book ANTO (Austrian National Tourist Office) or SNTO (Swiss National Tourist Office), and these are subdivided into areas – in Austria there are 9 tourist areas, one for each province, while in Switzerland there are 11, in some cases covering one canton, in other cases as many as four. (For tourism, the independent principality of Liechtenstein is part of the area of eastern Switzerland.) To arrange the land into a smaller number of regions with some distinct character, we have split it up as follows: **Vienna**, city of song, of waltzes, of gaiety, gladly burdened with the monuments of the Habsburgs, former melting-pot of the Empire, once again the crossroads of Europe. **Eastern Austria**, the provinces of Burgenland and Niederösterreich, probably the least visited part, with Alps, the Vienna Woods, the most beautiful stretch of the Blue Danube, and the steppes of the central European plain, all within easy reach of Vienna. **Southern Austria**, provinces of Kärnten, Osttirol and the green Steiermark with the relatively undeveloped and heavily forested southern Alps and the warmth of the southern lakes. **Northern Austria**, provinces of Salzburg and Oberösterreich, centred on the Salzkammergut, land of the White Horse Inn and Sound of Music, plus the musical city of Salzburg. **Western Austria**, the province of Tirol, which is the archetypal Austria of snowy, cow-covered Alps, dancers in Lederhosen, skiing and brass bands, plus the little province of Vorarlberg which is a sort of Swiss Tirol, and the tiny principality of Liechtenstein, a sort of Austrian canton of Switzerland. **Southern Switzerland**, the dry canton of Valais and the high canton of Graubünden, both winter sports grounds, and, between them, the sunny canton of Ticino. **Heartland Switzerland**, (Swiss tourist regions of Berner Oberland and Central Switzerland), the original centre of alpine tourism – William Tell country around Lake of Luzern and classic mountain scenery around the Jungfrau. **Northern Switzerland**, the flattish, heavily populated lowlands between the Rhine and the Alps, plus the pre-alpine pastoral hills of eastern Switzerland. **Western Switzerland**, French-speaking – the long mountains of the Jura, the agricultural part of the Swiss plain, and the resorts round Lake Geneva.

In its natural resources Switzerland is a very poor country and 50 years ago was a cheap holiday escape for the wealthy English middle classes. Yet today it has the highest national product per capita of any country in the world. (16,400 dollars compared with 9300 for Britain and 11,800 for USA). How they have done it, and what it has cost them are questions to ask yourself as you travel around.

Swiss wealth comes from manufacturing in a few selected areas calling for a high input of skill. Switzerland's biggest company is Nestlé, the food manufacturer, and the next five companies are three pharmaceutical producers and two machinery makers. Austria's principal product until fairly recently was timber, but manufactured goods based on her rich reserves of iron ore are now far more important, and even tourism brings in more money than wood. Despite the importance of industry to modern Swiss life (only 7% of the work force is engaged in agriculture or forestry), the Swiss picture of themselves as descendants of the tough, independent men of the mountains still lingers; the Austrians see themselves as descendants of the same stock, but with the culture of Empire.

The Swiss A lot of Swiss characteristics can be attributed to their will to get things correct and precise. It's as though, in taming the mountains, they've tamed their own nature. They subordinate themselves to the rules in order to profit from them. Bus drivers have always been punctual, but now they have electronic clocks which show the second, the driver delights to sit with hand on the brake, foot on the pedal, and eye on the clock waiting for the precise second to move off – even if he stops ten metres up the road for a latecomer. The law is for the good of everybody and it is for the sake of correctness, not out of civic duty, that a Swiss will inform on a neighbour infringing the planning rules. They expect value for money, and expect to give value for money. 'Keep the change' is not a Swiss concept – if a price has been set, then the exact price must be paid. This seems cold, but the Swiss can be warm once you're in the circle and show that you are playing the game by the same rules as they.

They may seem reserved and inward-looking until you know them, for nobody in the balanced Swiss world can be conspicuous. There is almost no ostentation in Switzerland – the Austrian touch of style, handed down from aristocratic days, is seen as snobbery by the Swiss.

The Austrians Where the Swiss are humane, the Austrians are kind. Austria grew up as the leading part of an aristocratic empire whose collapse is within living memory. This has left a taste for style, a cheerful cynicism, because the empire had been crumbling for centuries, and an open nature, because there was continuous movement of the subject people of the empire, always a newcomer to be welcomed.

In imperial days, Austria was not noted for efficiency or hard work – indeed, the opposites were virtues, the answer to an inefficiently oppressive aristocracy. For this, they have been called the 'Irishmen of Europe' (whether this is a slander on Ireland or Austria is a moot point). The maxim is still used: 'In England everything is permitted that is not expressly forbidden, in Germany everything is forbidden that is not expressly allowed, in Austria everything is allowed that is expressly forbidden'.

Things have changed and are changing fast now that the empire, the Germans and the Allies have all gone. The new Austrians have worked hard to build a country out of nothing, and continue to enjoy work. The law is obeyed, and the trains leave on time; more and more, the characteristics associated with the Swiss are becoming Austrian. But Austrians still make you feel that work and precision are unfortunate necessities of a wicked world, and that *Gemütlichkeit* will still break out.

Both countries being small and neutral, are free of the tensions that you may feel in a larger nation that believes it has a world role to fulfil.

THE ARTS

A tour round the birthplaces and working homes of the greatest masters of music is a fine way for the music-lover to visit Austria. Here are some of them: **Gluck** (1714–87), born at Neumarkt, spent his life in Vienna where he created opera in its modern form, leaving Vienna, even today, the supreme home of opera. **Haydn** (1732–1809) lived and worked at Eisenstadt (p. 48) as court composer to the Esterhazys, and created the string quartet in preparation for **Mozart** (1756–91), born in Salzburg where his house is a centre of pilgrimage for visitors to the great Mozart festival (p. 67). **Beethoven** (1770–1827), born in Germany, came to Vienna as a young man and his greatest carol *Silent Night* at Hallein (p. 62), where he lies buried. Johann **Strauss** the elder (1804–49) created the Viennese waltz, his son Johann (1825–99) wrote more great waltzes and created the operetta. He was followed by **Lehár**, whose operetta symbolizes past Vienna – gay, sentimental, superficially frothy. The break-up of the Empire is perhaps paralleled by the exploration of atonal music developed by the Austrian school of the first half of this century – **Berg**, **Schönberg** and **Webern**. Of the greatest names of European music, only the immortal Bach was

Mozart, Burggarten, Vienna

Schubert, Stadtpark

Beethoven, Beethovenplatz

Strauss, Stadtpark

works all received their first performance there. See Heiligenstadt (p. 45). **Schubert** (1797–1828) was a total Viennese, and the café life in which he wrote songs for his drinking companions can still be faintly felt; his lyricism embodies Vienna. **Bruckner** (1824–96) wrote his symphonies in Linz (p. 65) and was followed as a symphonist by **Mahler** (1860–1911) in Vienna. **Liszt** (1811–86) was born at Raiding in Hungary (now in Austria, p. 48) and is recalled at the Haydn museum in Eisenstadt. **Gruber** (1787–1863) composed the Christmas not part of the Austrian scene – even **Brahms** preferred to live in Vienna. This rich musical past is reflected in modern Austria, where you can rely on hearing a decent orchestra frequently in any town of size.

Honegger (1892–1955) apart, Switzerland has not ventured into the big league of composers – as with other cultural achievements, the Swiss art is really their way of life; but musical culture is strong in Switzerland, with an established, busy and unsubsidized orchestra in each main city.

Austria's other great contribution is in architecture. After the Thirty Years' War for the re-establishment of Catholicism in Austria, and especially after the final defeat of the Turks in 1683, the flowery Baroque style was taken over from Italy, and the churches, abbeys and palaces in which Austria still abounds were built or redecorated with sumptuous embellishment which should be seen even if you don't like it. **Fischer von Erlach** and **Von Hildebrandt** are the two architects most favoured, whose larger work such as the Trinity Church in Salzburg is saved by restraint. In Switzerland, only St Gallen, Disentis and Einsiedeln abbeys show real Baroque, elsewhere the style is much more muted in keeping with the temperament of the people.

After the Baroque period there developed the style which the Austrians call 'classical' – from about the middle of the 18th century – very similar in its pediments, ornament and proportions to English Georgian of the same period. It is remarkable that two such similar architectural styles should develop in England and Austria at about the same time quite independently of each other. To an English eye, Austrian classical looks less harmonious than Georgian, for it grew out of a less harmonious society, and the Austrians were fond of plastering their buildings in Maria Theresia ochre in place of Georgian stock-red brick. While there are only residues of Georgian left in England, the Danube basin, Steiermark and Kärnten contain whole towns with a complete classical heart. These are on the same pattern – a broad main street, usually closed at each end by an archway, with a town hall and a column or fountain commemorating escape from the plague somewhere in the middle. There is a variety of styles from late medieval up to classical, since when time has stood still, so that the heart of the town looks much as it did 150 years ago. Switzerland has its old towns, too, with a medieval heart whose charm is in the group as a whole rather than in individual buildings. A special Swiss contribution is the arcades (*Lauben*) which give at the same time intimacy and grandeur to the urban scene; the best examples are in Bern. Viennese architects at the start of the 20th century led the way in less flamboyant modern styles; the Swiss **Le Corbusier** followed.

In painting and sculpture, the Alpine countries have not come into the front rank; the Swiss **Paul Klee** is probably the artist best known among non-specialists. In literature, too, there are few internationally famous names, because Swiss writers – even Rousseau – have tended to look to a larger language group in France, Italy or Germany, while the Habsburg empire never had a national language. Austrian literature of the turn of the century, reflecting the light-hearted cynicism and cheerful despair of life under a disintegrating social order, deserves to be better known – **Kafka**, of course, though he lived in Prague, novelist, playwright and poet **Stefan Zweig** and the novelist **Robert Musil**.

Freud in Vienna and **Jung** in Zürich founded psychoanalysis. The tranquillizers produced in Basel are a more modern solution.

In engineering, the Swiss have made a mark out of all proportion to their numbers, with their experience in road building, tunnelling, hydroelectric power and locomotives; the railway engineering of the Empire was as dramatic, and modern Austria gets world-wide contracts for civil engineering and steelworks.

But there's one art for which the Alpine countries are most justly famous – the art of living together. The Austrians can show you how to enjoy life, with their love of *Gemütlichkeit*. The Swiss are divided by language more deeply than Belgium, divided between Catholicism and Protestantism, divided between once-imperial and once-conquered lands, they've strong trades unions in the very haven of capitalism, there is the geographical conflict between the mountains and the plateau, and despite such divisions, everybody sacrifices self-interest in order to benefit from the common interest. Reserved and undemonstrative, they too offer *Gemütlichkeit*.

PAPERWORK

A passport is needed by any foreigner entering Austria or Switzerland; a British Visitor's Passport is sufficient, provided the first page is marked 'British Subject' or 'British protected person', and young people travelling on a British Collective Passport are admitted. Entry on a passport permits a stay of up to three months; for a longer stay you must report to the Aliens' Police in Switzerland or to the Police in Austria. You may not enter Switzerland for the purpose of employment without a written contract of employment and a permit to reside in the commune, which is obtained by the employer.

No visa is needed for holders of a valid passport from (*inter alia*) Australia, Canada, Irish Republic, New Zealand, United States of America, United Kingdom, nor do these visitors need a visa to travel through any EEC country to get to Austria or Switzerland. Visa needed for South Africans into Austria, but not into Switzerland.

There is no requirement for vaccination or inoculation for travellers entering Austria or Switzerland from European countries or the Western Hemisphere.

There is no state medical service in Switzerland and all medical treatment must be paid for. Medical insurance is therefore strongly advisable. In Austria, there are reciprocal arrangements between the Austrian and the British national health services, by which employed persons (not self-employed) and pensioners of the UK can get free medical treatment, but this requires filling in form E111 from the Department of Health and Social Security before you leave. Medical insurance is therefore advisable for Austria. Skiers should not need to be reminded to take out medical insurance. Your existing insurer will extend cover fairly cheaply, or a travel agent will have proposal forms ready, or you can get insurance from: International Association for Medical Assistance to Travellers, Suite 5620, 350 5th Ave, New York or 1268 St Clair Ave West, Toronto, or from Europ Assistance Ltd, 252 High St, Croydon, Surrey, England (UK residents only).

For first-aid treatment, visit a chemist – there will probably be no charge in Austria, and may well be no charge in Switzerland.

Insurance of property is sensible; loss or theft should be reported to the police and a written note of the report obtained to satisfy your insurers.

CUSTOMS

The duty-free allowances for alcohol and tobacco goods, which apply only to persons of 17 years of age and over, are:

Non-residents entering Austria or Switzerland

From a European country: Alcohol, 1 litre Spirits (*over 25°*), 2 litres Wine; Tobacco, 200 Cigarettes or 50 Cigars or 250gm Tobacco.

From a non-European country: Alcohol, 2 litres Spirits (*over 25°*), 4 litres Wine; Tobacco, 400 Cigarettes or 100 Cigars or 500gm Tobacco.

Persons entering Austria from the duty-free zone of Samnaun, or Switzerland from the duty-free zones of Livigno and Luino, may be granted less generous duty-free allowances – leaflets detailing the allowances are available at frontier towns.

Duty-free allowances of normally dutiable goods on returning home are:

Duty-free allowances *subject to change*			
		Goods bought in a duty-free shop	Goods bought in EEC
Tobacco (Double if you live outside Europe)	Cigarettes or	200	300
	Cigars *small* or	100	150
	Cigars *large* or	50	75
	Pipe tobacco	250 gm	400 gm
Alcohol	Spirits *over 38.8° proof* or	1 litre	1½ litres
	Fortified or sparkling wine plus	2 litres	3 litres
	Table wine	2 litres	4 litres
Perfume		50 gm	75 gm
Toilet water		250 cc	375 cc
Other goods		£28	£120
US customs permit duty-free $300 retail value of purchases per person, 1 quart of liquor per person over 21, and 100 cigars per person.			

Normal personal effects (eg clothing, sports gear, photographic equipment, camping gear) may be imported by foreigners into Austria or Switzerland free of duty. Switzerland has the odd restriction that food up to the normal requirement for one day only may be imported, and only 125gm of butter. There are special regulations on the import of meat into Switzerland. Firearms must be accompanied by a certificate from the Swiss or Austrian consul.

CURRENCY

The Austrian unit of currency is the Schilling (ÖS internationally, just S in Austria) divided into 100 Groschen. There are coins of 50 Groschen ($\frac{1}{2}$ Schilling) and 1, 5 and 10 Schilling; coins less than 50 Groschen used to circulate and may still be found but they are worth so little they have almost vanished. Notes are issued in denominations of 20, 50, 100, 500 and 1000 Schilling.

The German mark (DM) is almost a second currency in Austria and many prices are quoted both in ÖS and in DM (100ÖS/30DM). Hotel and restaurant bills, entrance fees, even speeding fines, can usually be paid in DM. Pounds sterling and US dollars, and travellers' checks in these currencies, are best changed in a bank or exchange office (passport needed for travellers' checks) but may be accepted in larger hotels and shops. Canadian, Australian and Irish notes are less familiar and should be changed in banks.

Austria has known no less than six different currencies in the last hundred years, whereas the Swiss franc (Sfr) has become the symbol of stability. The franc is divided into 100 units called centimes in French Switzerland and Rappen in German. There are coins of 50 centimes ($\frac{1}{2}$ franc), 1, 2 and 5 francs in circulation. Notes are of 10, 20, 50, 100, 500 and 1000 francs. Coins of 5, 10 and 20 centimes are still around but slowly disappearing because they are worth so little. 'All the world loves the Swiss franc, but it takes the Swiss to love the centime.'

Foreign currency and travellers' checks are best changed at a bank (passport needed for travellers' checks); pounds, dollars and marks are often accepted in payment but at a discount.

Credit cards of all the internationally recognized issuing houses are accepted in Austria and Switzerland in banks, major hotels, car rental firms. British travellers wishing to cash personal cheques will need a special Eurocheque card supplied by their own banks.

Banks in Switzerland in general are open Mon to Fri from 0900 to 1600 (with cantonal variations). In Austria the normal hours are 0800 to 1230 and 1330 to 1500 (1730 Thurs), but there may be local variations – for example, banks close on Fri afternoons in Oberösterreich, the banks stay open late on market day in some towns, or open late on Mons to no clear pattern. There are exchange offices open nearly 24 hours a day every day at the railway stations in Vienna, Salzburg, Linz, Villach and Graz, and exchange offices open about 18 hours a day at major Swiss railway stations. Exchange rates in Austrian and Swiss banks are the same everywhere for the day.

You may take out of Austria no more than 15,000 ÖS in Austrian currency but otherwise there is no limit on the amount of foreign, Swiss or Austrian currency you may take into or out of either Austria or Switzerland.

HOW TO GET THERE

By Air The national carriers are Swissair, a privately owned company, and Austrian Airlines, which is government supported. Swissair has created a legend as the world's most efficient carrier and tends to stand aside from the fares war. Austrian Airlines (the Friendly Airline) has become similar to Swissair. Most other international carriers fly into both countries.

The airports for scheduled flights from North America or UK are Vienna (Schwechat airport, about 30 mins from city centre); Zurich (Kloten airport, by road, about 20 mins – 10 mins by train – from city centre); Geneva (Cointrin airport – 15 mins by trolley bus) and Basel (Mulhouse airport, in France, 30 mins by duty-free highway from city centre). In addition, there are scheduled flights by Dan-Air from London Gatwick to Bern. Graz, Linz, Klagenfurt, Salzburg and Innsbruck have other international connections, especially from Frankfurt.

Any travel agent or airline will help you with flights to Austria or Switzerland, and the holiday pages of newspapers are packed with advertisements for cheap

Currency 15

16 *How to Get There*

flights. A package tour (flight plus accommodation) is normally cheaper than buying the ingredients separately.
By Road If you are touring Europe by car, then any road that crosses the Swiss or Austrian frontier is an entry point.

If you are driving direct from Britain, the shortest distance is from Calais to Basel, which can be cut down to 693km/433mi using D roads through France; for the shortest time – using motorways via Reims and Metz – Calais to Basel is about 760km/475mi. From Calais to Vienna via Nürnberg is 1310km/820mi, but there are many alternative routes which are little longer. The best channel crossing, for balance of speed and comfort, is by British Rail from Dover to Calais; other channel crossings are: Newhaven-Dieppe (Sealink); Folkestone-Boulogne (Sealink); Folkestone-Calais (Sealink); Dover-Boulogne (Sealink and P.&O.); Dover-Calais (Sealink and Townsend ferries, Seaspeed hovercraft); Ramsgate-Calais (Hoverlloyd hovercraft); Dover-Dunkirk (Sealink); Dover-Ostend (Sealink, good for Austria); Sheerness-Vlissingen (Olau, slow but good for Austria); Hull-Rotterdam or Zeebrugge. From Ireland, ferry from Rosslare to Le Havre. Complete collection of ferry operators' brochures at the AA or RAC in London.

There are direct coach services from London to Geneva, Zürich (20 hours) and Vienna (28 hours) by Europabus (Victoria Coach Station), and by up to a dozen private operators who can be found around Shaftesbury Avenue, London. Amsterdam-Luzern by bus is gruelling but cheap. As air fares rise, ski-tour operators are turning to coaches which manage British channel port to Tirol resort in 18 hours.

Cars can be taken by train from Calais to Lyss or from Paris to Evian, for Bern and Geneva respectively, and from Brussels to Salzburg and Villach and from Köln to Vienna.
By Rail Switzerland is the crossroads of Europe (especially Zürich, which at peak times can boast nearly 1000 international connections per day) and it would be futile to try to summarize the routes. Austria is nearly as well connected and Vienna, Salzburg and Innsbruck are especially well served by routes from Germany. From London to Basel is 13 hours by train (Edelweiss express), London to Vienna is 23 hours by train (Ostend express). Vienna to Zürich is 8 hours by train (5 daily). It is advisable to book a sleeper or couchettes if travelling by train.
By Ship A Rhine steamer from Amsterdam to Basel takes $4\frac{1}{2}$ days ($3\frac{1}{2}$ days back, *ie* downstream) and can be joined anywhere along the Rhine, a fun trip. Danube steamer from Passau to Vienna, enjoyable, but less dramatic than the Rhine. Lake steamers on Lac Léman from France, Lago Maggiore from Italy, Bodensee from Germany.

INTERNAL TRAVEL

By Air There are frequent direct connections, by Swissair and Austrian Airlines, between Zürich and Vienna.

Domestic flights are operated by Austrian Air Services between Salzburg-Linz-Vienna and Klagenfurt-Graz-Vienna; by Tirolean Air from Innsbruck to Zürich, to Vienna and Frankfurt; by Swissair to connect Geneva, Basel, Bern and Zürich; and by Alpine Lufttransport AG (ALAG) to connect Swiss mountain resorts to cities.
By Train Switzerland has a dense, frequent and punctual network of trains by which it is possible to get nearly everywhere (and places not served by train can be reached by buses timed to connect with the trains). The Austrian network has become equally dense in recent years, and equally efficient. The main lines in Switzerland are operated by the state railway (Schweizerische Bundes Bahn/Chemins de Fer Fédéraux/Ferrovie Federali Svizzerie) – you can recognize them by the engines, with a lovely cowling like a medieval Swiss helmet. Most mountain railways are privately owned – however the state and private railways charge the same fares, dovetail neatly, and there is a single timetable for the whole country. In Austria, too, the railways (Österreichische Bundesbahn) are state run.

The railway systems in the mountain districts of both countries are fascinating miracles of engineering, and railway enthusiasts from all over the world come to spend a holiday just travelling round by train. Trains are nearly all electric with a few diesels, *eg* in the Zillertal to Mayrhofen. As a tourist attraction, there are twelve lines operating steam trains in Switzerland and about eight in Austria. Traction is mainly by friction, but on the steep sections trains engage rack-and-pinion without passengers noticing.

Dining- or snack-car, plus drinks trolley are found on most main-line trains and telephone in most ÖBB inter-city expresses. You get on and off the platforms without ticket inspections; tickets are examined on the train by a conductor who comes round between each two stops; there is a small surcharge if you buy your ticket on the train instead of at the booking office.

By Bus Town buses and trams are run by the municipal authorities, in most cases you buy tickets from a machine at the stop before you get on. The Geneva system, whereby you buy a ticket for an hour's unlimited travel as you start the journey, is becoming the most popular. Once on the bus, there are spot checks on tickets, and instant fines for travellers without; it is no disgrace to travel without a ticket but averaged over many journeys it is twice as expensive. Travel in towns is cheap.

Country buses in Switzerland are run by the postal administration and usually called *die Post* or *la postale*; they serve places not reached by train. In Austria some ÖBB buses run parallel to the trains but at different times, but most buses are run by the postal administration, and are called *die Post*. There are some private buses in Austria, too. The times of all Swiss country buses are listed together with train and boat times in the *Amtliches Kursbuch*, but there is no consolidated timetable for Austrian buses; the network is dense and frequent.

The country buses in both countries are painted a brilliant 'post office' yellow. They are comfortable, meticulously clean, maddeningly punctual, fairly slow, and their reputation is that they always get through. They give a good view, and often stop for a few minutes at favourite viewing points.

By Ship Most lakes of any size have privately operated pleasure steamers; on the bigger lakes – Léman, Neuchâtel, Luzern especially, and the Bodensee – there is a regular and frequent service which is useful as a means of getting about as well as providing fun outings. Details in the Swiss *Amtliches Kursbuch* blue pages. Ships on the smaller Austrian lakes are just for outings.

Travel on the Danube is by Donau Dampfschiffsfahrtges. (DDSG), based at 8 Mexicoplatz, Vienna (by the main Danube quay). The most favoured stretch of the Danube for pleasure outings is through the Wachau, between Melk and Krems.

Reductions Children up to 6 travel free on trains and buses if accompanied by a fare-paying adult. Children aged 6-15 (in Austria) or 6-16 (in Switzerland) pay half fare, this reduction extends to any other concessions listed below. Eurailpass is valid in both countries.

For extensive travel, good value are the Swiss Holiday Card which gives unlimited free travel on trains, country buses, many lake boats, and some reduction on mountain transport and other boats, and the Austria Ticket which gives free travel on trains and buses of the ÖBB and Post, and Wolfgangsee boats, and about 50% reduction on private (mountain) railways, private buses, Bodensee and Danube boats. Swiss Holiday Cards are available for 4, 8, 15 or 31 days, Austria Ticket for 9, 16 or 31 days; both, for non-residents only, from National Tourist Offices. There are variations on these cards, eg available for one province or one tourist region, but on the whole it is harder to get your money's worth from these cheaper but restricted cards than from the national cards. With your Swiss Holiday Card, get the official timetable, which gives the times and distances for every train, bus (except city buses), boat, and cableway, public or private, in Switzerland – a marvel of compression. The Austria Junior ticket gives a two-thirds reduction on trains, many buses, some funiculars and shipping lines – available to persons under 23. A similar ticket is available for persons over 65. The Swiss Half Fare Travel Card, which is cheaper than the Swiss Holiday Card, allows you to buy tickets for nearly all travel in Switzerland, except cable cars, at half price. For city travel, you can get a book of ten tickets, each valid for one day in any one of about twenty towns (bought from main offices of the bus company, usually by the main station). A return ticket in Austria is about 80% of the price of two singles. Switzerland gives a reduction of 3–18% for journeys up to 64km/40mi and 20% for journeys of 65km or more, for return journeys.

Nearly every resort offers some sort of facility card to guests staying a week or more, and sometimes for only three days, giving worthwhile reductions or free use of facilities – swimming pools, tennis courts, and those overworked ski lifts.

By Bicycle Bicycles are carried on Swiss and Austrian railways for a nominal charge if they go on the train with the passenger. Bicycles can be rented very cheaply at most Swiss railway stations (619 of them) and about 10 Austrian, and left at any other station in the country of rental. If you use trains for the heavy work, there are large stretches of good cycling country even in the mountains.

Hitchhiking is illegal on motorways, otherwise permitted and fairly easy.

Mountain Transport There are five main types of conveyance for taking you up a mountain to a viewing point, and their names are sometimes confusing. A funicular is a pair of cars, which run on rails and are drawn by cables, one car goes up to balance the other car coming down; 100 or so passengers in each car, called *Standseilbahn* or *funiculaire*. A cable car can be the same size as a funicular, but is suspended from a cable high over the valley: *Luftseilbahn* or *télépherique*. Funiculars and cable cars run to a set timetable, say every 15 minutes.

A gondola is a much smaller car, for 2–6 people, also suspended from a cable. The cable runs continuously, and the gondola is hitched on to it once the passengers are inside: *Gondelbahn* or *télécabine*. A cable chair is a seat, unenclosed, which like the gondola is hitched to the cable once the passenger is on it: *Sesselbahn, Sessellift* or *télésiège*. All four of these run in summer for sightseers and in winter to take skiers up to a starting point; there are nearly 500 in Austria and over 200 in Switzerland. In addition, there are ski lifts (T-bars or drag bars), where a bar hangs down from a cable which the skier holds on to while keeping his feet on the ground: *Schlepplift* or *téléski*. These operate only in snow (about 2800 in Austria, 1500 in Switzerland).

Warnings It can feel very cold at the top of a mountain, even in bright summer sunshine, if you are suddenly transported there from a warm valley, so carry extra clothing. All mountain transport is expensive. People with a heart condition should not go above 2000m/6500ft.

Taxis There are taxis in all main towns, particularly at railway stations. Fares are shown on the meters. For journeys out of town, establish the fare beforehand, the meter charge can be astronomical. Village taxis usually have set rates for a small number of standard journeys. In Austria a tip of 10% is expected; in Switzerland the tip may be included, in which case the odd franc is accepted, otherwise 10% is sufficient.

Car Rental The major car rental firms have offices at the international airports and in main towns. Rates and conditions are roughly the same as anywhere else in western Europe; car rental is restricted to persons over the age of 21. The airlines have deals with the rental companies so you can book a car when reserving your flight; both Swiss and Austrian railways have similar deals with car rental firms, but this usually entails returning the car to the station where you picked it up.

Smaller garages also rent cars – you can find them in the phone directory under *Autovermietung* or *Location de Voitures* or you may spot them in passing. If they are just agents for the big firms they are no cheaper, but if they rent their own cars they can be at least one third cheaper. Locally rented cars have to be returned to their base.

IF YOU ARE MOTORING

Documents A foreigner driving a car or motorcycle in Austria or Switzerland must be at least 18 years old, and in possession of a full, valid driving licence issued by his country or state of origin. No translation of the driving licence is needed. A British provisional licence is not valid.

No customs document or carnet is needed to bring a private motor vehicle or trailer into Austria or Switzerland. It is obligatory in both countries to have third-party insurance; the vehicle insurance issued by a UK or Irish insurer automatically gives the minimum cover required by law in Austria and Switzerland, so no Green Card is needed. But for comprehensive cover you must get a Green Card from your insurer. The police can require you to produce the vehicle registration document and the insurance certificate in the event of an accident, so take them with you.

The car and any trailer must carry a sticker showing country of origin (*eg* a GB plate).

Rules of the Road

Drive on the right, overtake on the left. Give priority to vehicles coming from the right when the road itself does not show which is the major and which the minor road (major roads have priority); trams have priority over cars; cars may not overtake a bus which is indicating that it is about to move off (this rule is ignored by nearly every Swiss or Austrian driver). Postbuses have priority over cars on mountain roads. It is obligatory for the driver and front passenger to wear a seat belt and for motorcyclists to wear a crash helmet. It is illegal to cross a continuous white line, or to park where you would force other traffic to do so. Vehicles going up have priority over those (except buses) coming down.

Lights Dipped headlights must be switched on in a tunnel, even if the tunnel is lit. In poor visibility (dark, snow, fog or heavy rain) dipped headlights or fog lights must be switched on, not just side lights. Motorcycles must, in Austria, have lights on at all times.

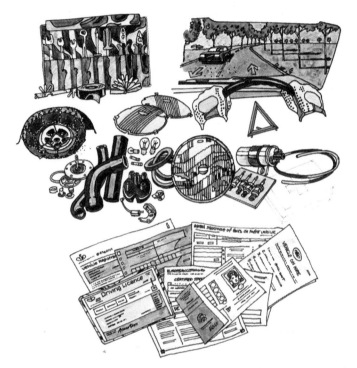

Speed limits The speed limit on motorways is 130kph/81mph; on other roads 100kph/62mph; in towns it is 60kph/37mph in Switzerland and 50kph/31mph in Austria. These limits are enforced by the police who impose on-the-spot fines. For a car towing a trailer the overall speed limit is 80kph/50mph, and in Austria if the trailer weighs more than 750kg the limit is 60kph/37mph, while in Switzerland, if the trailer weighs more than 1000kg the limit is 60kph/37mph. Advisory speed limits are marked on dangerous stretches and on the approaches to towns.

Road conditions The roads of Austria and Switzerland have in recent times become almost too good – many delightful little places, that even ten years ago the motorist would have come across while crawling from here to there, are now bypassed and you can get across the mountains almost as easily as across a plain, and far more interestingly. The motorway system is nearly complete, and includes some spectacular pieces of engineering, and tunnels are constantly being bored to make dramatic new connections or just to short-circuit a difficult little stretch. In general motorways and tunnels are free, but in Austria a toll (*Maut*) has to be paid on some stretches of sensational road building (*eg* the Brenner motorway and the road over the Tauern and Grossglockner and under the Arlberg). Traffic is not too heavy and keeps moving – the main holdups are on Sunday evenings when town dwellers return from a weekend in the country and at places where the relentless road-improvement machinery is at work.

Nevertheless, there is still a huge number of mountain roads on which you should leave the fast driving to the locals who presumably know what they are doing. In particular, select an intermediate gear going downhill, and leave the brake pedal alone.

In winter snow must be expected at any time and despite industrious salting and snowploughing it takes only one metre of blocked road to hold you up for hours. It is stupidity to motor in Austria or Switzerland in winter without taking snow chains. On most passes signs are put up when chains are advisable, and chains are often obligatory. Snow chains can be rented at filling stations, and sometimes on one side of a pass to be dropped at the other. Many passes are closed by snow in winter; most maps show the normal dates of closure, *eg* Oct to May, but there can be snow from early Sept to June, so watch for signs. The gradients on a mountain pass are usually quite gentle.

In Switzerland, the Klausen pass is forbidden to trailers, and the Albula, Bernina, Furka, Grimsel and Umbrail passes are 'not suitable' for trailers: not suitable means that if, as is likely, you can't manage it, you may be fined. In Austria, the Pass Gschütt, Seeberg Pass and Turracher Höhe defeat most cars with trailers. In general, don't attempt the alpine passes with a trailer unless you can manage a 1-in-4 hill start and can reverse round a bend downhill (a manoeuvre more fun to watch than to attempt in snow at night).

Petrol (gasoline) Two grades of motor spirit are available – normal, which is 92 ROZ, equal to British 2-star, and super, which is 97 or 98 ROZ, equal to British 4-star. Some pumps can provide blends. *Benzin* or *Essence* does not mean all grades of fuel, but just normal grade. Fuel is marginally cheaper in Switzerland than in Austria. In each country, fuel is about the same price everywhere – no dearer on motorways or at remote filling stations; there are discount filling stations round some towns, but the suppliers are trying to stamp out cut-price fuel. Filling stations open at about 0700 and start closing between 1900 and 2100, earlier in winter; there are plenty of all-night filling stations on main roads.

Alcohol The police use the breathalyzer, as elsewhere, to detect excess alcohol in a driver. The limit in Austria is 80mg/ml (the same as in Britain) and the penalty is 5000 ÖS and confiscation of driving licence. The limit in Switzerland is also 80mg/ml and the penalty depends on the severity of the offence.

Breakdowns The law requires that you carry a reflective triangle for use as a warning if you break down and, in Austria, a first-aid kit. If you belong to a national motoring organization, get the certificate of the Alliance Internationale de Tourisme, which gives you the benefits of membership of the local organizations. These are the Touring Club Suisse, Automobil Club der Schweiz, and the ÖAMTC (Österreichischer Aotomobil-Motorrad-und Touring Club) and ARBÖ (Auto-Motor-und Radfahrerbund Österreichs). Their help to get you going or to a repair garage is then free or for a small fee. They can be contacted by telephones spaced along the motorways, or any Gasthaus bar or even a private house with a phone will summon help for you. (A breakdown is *Panne* in both French and German.)

Dealers carrying spares for British-made cars are scarcer than in Britain, but sufficient to let you get spares in all main towns. The motoring organizations rent out a kit of basic spares which you can return at the end of your stay, if unused.

Parking On mountain roads, the best views seem always to be at a bend, but stop well away from the bend, pull off the road and walk back. If you create a dangerous situation by thoughtless parking, you can be fined.

More and more towns are creating traffic-free inner cities, which makes parking more difficult. There's usually good parking on the outskirts – use that and stretch your legs. Street parking is quite disciplined everywhere; some on parking meters, more on time-limit zones where you can leave your car for a set time, usually 1½ hours (once you find a space) but must display the time of arrival on a dial card (*Parkscheibe* or *disque de stationnement*). Parking is forbidden on tram routes.

Horn On mountain roads, it used to be normal practice to sound the horn on bends; this precaution is dying out, even with the postbuses, but occasions no comment if you do sound the horn. Use of the horn is forbidden in towns at night, and in Vienna at all times.

ACCOMMODATION

Hotels Switzerland can claim to be the birthplace of inn-keeping for travellers, and it is the home of modern hotel-keeping for holidaymakers. The standard of cleanliness and efficiency, and the right blend of hospitable warmth and discreet reserve, are unique. Austria, starting from a different base, has almost caught up, and if there is the occasional lapse it is offset by friendliness. Both Austria and Switzerland have a stock of older, grander hotels built either for the moneyed classes of the Empire or for the aristocratic Englishman, and these have been modernized – showers, *etc* – with a certain loss of spaciousness. The impersonally efficient style of modern American hotel-keeping has not taken hold in either country.

Strictly speaking, 'Gasthof' is the equivalent of the old English inn, a place out in the country offering good company and a bed for the night, while a 'Hotel' is built as a place in town to go and stay. But the distinction has disappeared – an old Gasthof may be enlarged and enlarged until it thinks itself grand enough to be called a hotel, while a modern hotel may call itself a Gasthof for the folksy sound. We will call them all hotels. A hotel garni is one which serves only breakfast, no other meal.

Swiss hotels are graded from one star to five stars; the few one-star hotels in towns may be a bit rough, in the country they are primitive but homely. Two, three and four stars represent progressively more facilities, five stars represent luxury and commensurate expense. A three-star hotel is not much dearer than a two-star, and can give a good deal more in space and facilities; four stars is a marked step-up in price with a less noticeable improvement in facilities. The grading is done by the Swiss Hotels Association. A list of hotels for the whole of Switzerland, *Swiss Hotel Guide*, is published by the association (available from SNTO) and this is almost complete for three-star hotels and upwards. Each of the eleven Swiss tourist regions publishes a list of hotels in its region, which is complete for all hotels. Both lists give minimum and maximum prices for a room per day, and these prices are also posted inside the rooms. It is usually the minimum price which is charged.

In Austria, hotels are similarly graded on a scale from one to five stars. (The previous grading system, by letter, was phased out in 1982.) Grading is done by the Chamber of Commerce. Swiss hotels are about twice as expensive as Austrian in the lower grades, but the price difference lessens as you move up. A list of hotels covering the whole of Austria, *Hotel List Österreich*, is available from the ANTO. Some smaller hotels are not on this list. Each of the nine provinces publishes a list of hotels in its area, and these lists are complete. Both lists state minimum and maximum prices for a room for one day as reported by the hotel; again, it is usually the minimum price which is charged.

The stated prices are for bed and continental breakfast. A surcharge is permitted for a stay of one day only, but is seldom levied.

Very substantial reductions are given for a stay longer than three days in many hotels, if the booking is made in advance. These all-inclusive terms are called *Pauschalangebote* or *Pauschalpreise* or *prix à forfait*. Details of these reduced offers are available from the local tourist office of the resort you want to stay in.

In Switzerland, children up to 6 years old pay at the discretion of the hotel, children up to 12 pay 50% of the adult rate and children up to 16 pay 70% of the adult rate. In Austria children up to 4 pay 50% of the adult rate and up to 10, 75% of the adult rate. These reductions are for children in the same room as their parents, and are not a legal requirement.

There are about thirty castles or palaces in Austria which have been converted to four- or five-star hotels, they are shown in a booklet *Schlosshotels & Herrenhäuser* available from ANTO.

Motels There are just over 100 motels in Switzerland, listed in a leaflet *Motels* from SNTO. In Austria, there are some along motorways.

Farms Many farms take in 'paying guests'; sometimes this is a true farm with a spare room used to make a little money on the side, and sometimes the guest accommodation has been extended to make visitors the main source of income. These are found most readily in Austria. Details can be got in a leaflet *Ferien im Bauernhof* or *Urlaub auf dem Bauernhof* published by most of the provincial tourist offices (except of course Vienna).

Private homes In many villages you can find rooms in private houses. Prices are about the same as for a one-star hotel in the same district. You can recognize them by the sign *Zimmer Frei* in the window, or they are listed in leaflets published by the regional tourist offices. Lists and prices of private rooms are sometimes included in regional hotel lists, but not in national hotel lists.

Apartments Holiday chalets, bungalows, or separate apartments in houses give the opportunity for self-catering if your budget is tight. These are often abbreviated to 'Fewo': *Ferienwohnung*, or holiday dwelling. Costs vary widely according to space, location and furnishing, but think in terms of the cost for a family as the same as the cost for one person in a four-star hotel. Lists of names are available from Swiss regional tourist offices; more detailed lists, giving facilities but not always prices, are available from Austrian provincial tourist offices. Cost of heating and cleaning is usually extra.

Youth hostels There are just on 100 youth hostels in Austria, and 120 in Switzerland; they are open to young people (nominally up to 26 and 25 years old respectively) and to parents accompanying young people. Foreign youth hostellers must be members of their national youth hostel association affiliated to the international association. Location of hostels is shown in the IYHA handbook, or they can be found in the telephone book under *Jugendherberge* or *auberge de jeunesse*. For other cheap student accommodation consult the Schweizerischer Studentenreisedienst, 19 Leonhardstrasse, Zürich. There are university student hostels which operate as hotels in summer in Vienna, Linz, Graz, Klagenfurt and Salzburg.

Camping Austria and Switzerland are a camper's paradise; there are over 300 sites in Switzerland approved by the Swiss Camping and Caravanning Association, and over 400 in Austria. Nearly all the sites are small in comparison with the camping cities that have sprung up along the Mediterranean. For Austria, get the booklet *Camping – Caravanning* from the ANTO; for Switzerland get the map *Camping in Switzerland* from the SNTO.

In general, camping (with a big frame tent or a trailer) is not permitted except on established campsites, though if you can find the landowner you can camp wherever you are given permission. Camping with a bivvy tent in the mountains is quite acceptable.

Trailers (caravans) see If You Are Motoring (p. 20).

Spas There are 21 towns in Switzerland officially classified as spas and about 40 spas in Austria. Essentially a spa has spring water or muds containing salts with healing properties, but their important characteristic is the accompaniments to help you heal in comfort – restful parks and gardens, hotels on a rather grand scale where you can be pampered while 'taking the waters', gentle amusements all in a beautiful landscape. The search for healing properties need not be taken too seriously – a stay in a spa is an opportunity for complete relaxation which is probably the real benefit. There's always a cure centre with all sorts of medical treatment and exercise. The selection of a spa may depend on your doctor, different spas are for different ailments. There is little choice but to stay in a hotel, the cost of which may include attendance at the cure centre. Some spas, such as Badgastein and St Moritz Bad operate as spas in summer and ski centres in winter; some, such as Warmbad Villach and the Swiss Baden, remain spa-like despite the proximity of industry, others, such as the Austrian Baden and the once-great Bad Ischl, survive on the past. There has been a considerable upsurge in spa-visiting in recent years and package tours are replacing the former exclusiveness.

A place called a 'climatic resort' or 'health resort' has a physical atmosphere considered beneficial to health but the spiritual atmosphere is less intensely remedial.

Details of cures are published in booklets *Swiss spas* and *What is a climatic resort?* from SNTO, and *Nature heals* from ANTO.

FOOD AND DRINK

Where to get it There are few restrictions on the sale of alcohol in Austria, and none in Switzerland, so there is no sharp distinction between somewhere to eat and somewhere to drink. You can get a beer or a coffee in a dining hall with laid tables (preferably not in the middle of a busy lunch time) and often a simple meal in a *Bierstubli* or *Weinstubli* (beer or wine bar). A *pinte* is a bistro-type bar in French Switzerland; *Imbiss* is a snack bar. There's a huge range of eating places from station buffets to plush restaurants. At the bottom end of the scale, fast-food joints with an American flavour are proliferating in Switzerland, while in Austrian towns, kerbside stalls and street-sales counters of restaurants offer goulasch soup and spicy sausages until late into the night (but no alcohol after 2100); these are for convenience, but not for quality or cheapness. For pleasant snacks, amounting to a meal, many butchers' and fish shops in Austria, and bakeries in both countries, have tables and chairs during shop hours. The Swiss Migros chain of stores has long set the standard for cheap, good-quality, cafeteria-style food, and the Coop is now offering the same style; self-service restaurants attached to big stores are also attractively priced, though not so rock-bottom as Migros.

Moving up, a *Gasthaus* or *Gasthof* is good for a meal, as well as a glass or cup, and they are open from, say, 1130 to 2300; food will be simple, without much choice, but it will be prepared from good, basic ingredients and well cooked. During normal lunch hours these establishments will offer a *Tagesteller* or *assiette du jour*, consisting of soup and a cooked meat dish (no sweet), and this is the best buy if you are watching the pennies. Ordinary restaurants in towns offer wider choice; hours tend to be strict, *eg* 1200 to 1400, and 1800 onwards, and in Switzerland many of the less plush restaurants close at about 2100 – Austrian hours are later but in many cases not by much. In ski resorts in season and in main towns inexpensive places stay open later.

In hotels and the more expensive restaurants, the hours are more elastic and the food superb. The centre of the cuisine is international French, with regional specialities in Switzerland and specialities from the old Empire in Austria. The biggest danger is quantity. The habit in Switzerland is to serve only a half-portion to begin with – most Swiss, slim or fat, seem to manage the other half. Small portions for children are available at reduced prices. Cafés serve only drinks and light snacks, but you can sit on a terrace by the mountain or lake, or in the town square taking in the scene. For evening drinking, a 'bar' is essentially for alcohol, but a Bierstubli or Weinstubli, essentially for drinking beer or wine, may also serve coffee or a meal if you ask.

The *Kaffeehaus*, a speciality of Vienna, is slowly disappearing; here you can sit all day over a cup of coffee, reading the paper supplied in its wooden holder. Pastry shops are plentiful in Switzerland and the *Konditorei* is a special feature of Austrian life – *Jause*, the afternoon break from 1600 to 1700 for coffee and a rich cake, is a ritual similar to English tea.

The *Heuriger* (wineshop) is found in Austrian wine-producing districts. Strictly speaking, 'heuriger' means 'of today', and heuriger Wein means new wine. Wine is 'new' until 11 November of the year following the vintage, and the producer can sell his new wine from his own premises without a licence, for 300 days of the year. On the days he is selling, the grower puts up a bush or wreath as a sign and then is said to be *ausg'steckt*. The wine is served in the grower's own garden or courtyard either at individual tables or round a communal bench with singing and chatter to keep the flask moving. Customers can bring their own food. Many Heurigen provide an accompaniment of violin or accordion music and some provide food at a buffet. There are also so-called Heurigen in the middle of Vienna and even one in Zürich, but these are not necessarily the real thing. Heurigen are most frequented in the evening, and when the sun shines it's a gorgeous way to spend the afternoon.

Drink The **beer** is a lager type, mainly sold by the bottle but available on draught (*vom Fass*) in a Bierstubli and some cafés; mostly quite strongly hopped. About twelve breweries in the two countries, all good beer, none outstanding. A wide choice of **soft drinks**, with the fruit juices especially recommendable – apple juice from the southern provinces of Austria is noteworthy. Cola drinks and carbonated drinks are not so popular as in other countries. **Tap water** is of course clean, safe and refreshing everywhere. **Mineral water** from the many spas is readily available and quite inexpensive. **Tea** is made with tea bags but if you ask for it with an English accent it will come reasonably strong and with milk.

Coffee in Switzerland is usually real coffee in any place other than a station buffet, or, if instant, they manage to get it to taste like real. Served in quite small cups, with cream, except at breakfast when it comes in large cups, weaker and normally drunk black. In Austria, and above all in Vienna, coffee is a more developed art. A simple *Portion Kaffee* is a can of bean coffee with a jug of warm milk; at breakfast you may get *eine Melange*, a large cup of half-coffee half-milk; *einen Braunen* is mainly coffee with a dash of milk; *Mokka* is a small cup, strong, black and generally over-sweetened; *mit Schlag* is with a dollop of whipped cream (or *mit Doppelschlag* with two such dollops); *Türkischer Kaffee* is the Balkan black coffee soup.

A light, pleasant **wine**, mostly white, is produced in many areas of Switzerland and Austria, and it is the standard drink in western and southern Switzerland and eastern Austria. The specialities of most regions would be recommended as a particularly good buy, but there's not really much to choose – they're all drinkable, easy on the palate, unmemorable. Just take the wine of the district – usually offered as a carafe (*offener Wein*). For wine by the glass, the usual measure is a *deci* in Switzerland (0.1 litre), and an *Achtel* or *Viertel* ($\frac{1}{8}$ or $\frac{1}{4}$ litre) in Austria.

Spirits Switzerland produces a good variety of fruit-flavoured brandies – apple, plum, cherry as a special favourite, and pear. *Gentiane* is a distilled spirit flavoured with mountain herbs. Schnapps is popular in Austria; Austrian rum, which comes as 80% alcohol, is very cheap but not used for drinking, it's for cooking (makes pastry workable). Black coffee with rum is called *Mazagran*.

There are few hang-ups about drinking, so long as you don't mix it with driving. You can get a beer first thing in the morning, or get coffee or a soft drink in a place that looks like a drinking den, without pressure to take alcohol. Café, Gasthaus, Weinstubli, or *pinte*, they're there to provide the customer with what he wants.

Milk It may be worth bringing back a few bottles of *Alpenmilch* to put in your coffee at home to remind you of the flavour; most cow's milk contains 4% fat, but the milk from alpine cows has 7%.

Food The basic food style throughout nearly all of Austria and Switzerland – the sort of thing to expect if you get a meal in a country inn out of hours – is Germanic with added finesse. Pork chop with salad or sauerkraut and, as an alternative to ubiquitous French fries, either *Röstli* in

Switzerland or *Knödel* in Austria. *Röstli* are mashed potatoes pan-fried to a crisp brownness, *Knödel* are dumplings. But this factual account does not do justice to the high standard, almost loving care, of preparation and presentation. The food of both countries is like the Swiss of legend – clean, crisp but homely underneath.

Onto this basis is welded a great variety of regional styles derived from French or Italian Switzerland, or from the Hungarian, Bohemian and Slav territories of the Habsburg empire. Pizza and spaghetti, which are the norm in Ticino, are popular over the rest of Switzerland. French *haute cuisine* (with lighter sauces) is widespread in the west of Switzerland but found everywhere. Prague ham, *cevapcici* and *shashlik*, and a Hungarian spiciness in the goulasch soup, enliven Austrian cooking.

An important feature of Swiss cooking is the use of cheese – not only in fondue and *raclette*, but also in a great variety of sauces and tarts, such as the *ramequins* (eaten cold) of western Switzerland and the *Käsewähe* (hot) of the east. Fondue, which seems to be *de rigueur* for a ski party, is cheese melted in white wine in a communal pot from which you help yourself with a piece of bread on a long fork – the penalty for dropping your bread in is to buy another bottle of wine. Raclette is a wheel of cheese heated before a fire (or electric element) and scraped away as it starts to melt to form a crisp layer over potatoes. With *fondue bourguignonne*, you cook a piece of steak on the end of a fork in a bubbling pot of oil, wine and cheese. *Reblochon* is a creamy, soft, peasant cheese sold direct from the farm.

The great home of superbly rich, fluffy cakes and pastries is traditionally Austria; the famous Konditorei (pastry shops) still exist in Vienna and lesser ones all over the place, though Austrians themselves seem to have turned from cakes to set a greater emphasis on meat. Air-dried beef (*Bündnerfleisch*), cut very thin, is a feature of Swiss eating, and sausages are a feature in both countries (popular from a stall in Austria).

Breakfast, included in a bed-and-breakfast price, is just rolls with tea or coffee. A British-style cooked breakfast is available as an extra, but in package tour hotels, where the guests have to prepare for a day's strenuous exercise, a full breakfast is often included.

Soups tend to be heavy – barley and flour soups. Austrian *Goulaschsuppe* can vary from just strongly seasoned to burning; much milder is the Austrian clear soup with egg or with *Leberknödel* (liver dumplings).

ENJOY YOURSELF

The most prominent activities are naturally the skiing, mountaineering and alpine walking that attract the majority of visitors to these countries, but there are plenty of others. The simplest sources of information on where to enjoy such activities: for Switzerland, a summer sheet and a winter sheet entitled *Switzerland at a glance*, or duplicated sheets on individual sports, available from SNTO; for Austria, a booklet issued by each of the 9 provincial tourist offices, listing each sport with facilities and costs, available from ANTO.

Canoeing No restrictions on your own small canoe; very little rental. The river Enns, especially the torrents either side of the Gesäuse, is for experts; the Traun, the Inn below Mötz (Tirol) and around Celerina (Graubünden), the Doubs and the Ziller are all popular. In general, March to May is the time for streams (not in the high Alps), June to August is for the rivers fed by glacier water.

Caving There are over 20 caves in Austria with lighting and guided tours, the richest areas being the Dachstein and around Weiz in Steiermark. For independent caving, contact Association of Austrian Speleologists, Obere Donaustrasse 99/7, Vienna A-1020.

Climbing Real mountaineering is too serious to be condensed here – there is a saddening cemetery at Zermatt for British climbers. Plenty of climbing schools with qualified instructors. There are about 700 climbing huts in Austria and 200 in Switzerland. A few days' acclimatization needed before climbs of over 3000m/ 9750ft. Information from the Austrian Alpine Club or its British branch at Fretherne Rd, Welwyn Garden City, Herts, or the Swiss Alpine Club or its British section at 52 North St, Maracham, Oxford OX13 6NG.

Fishing To fish in Switzerland you need a cantonal permit, obtainable from the local police or in large towns from the tourist office, and a permit from the owner of the riparian rights, which is often the canton. In Austria you need an official licence (from provincial authorities) or a temporary authorization (*kurzfristige Fischer-Gastkarte*) which is usually available from the owner of the fishing rights. You also need a permit from the owner of the rights. There is virtually no rental of tackle. About eight varieties of coarse fish, and brook-, rainbow- and lake-trout. Many local restrictions on weight and bait. For serious fishing, get the booklet *Angelsport in Österreich*.

Enjoy Yourself

Golf There are 29 courses in Switzerland and 19 in Austria, mostly near the more expensive resorts. The scenery compensates for the fact that it doesn't feel like a native sport. Clubs can be rented. Mini-golf all over the place.

Hobby holidays (*Hobbykurse, cours de bricolage*) Holiday courses are available in specialized subjects including geology, ornithology (especially Lake Neusiedl and Zuoz), painting (many centres), pottery (very many centres), woodcarving (very many centres), cookery, winemaking, abbey visiting, botany, *etc*. Best listing of details in leaflets, arranged under province, from ANTO or, arranged under activity, from SNTO.

Horse-riding Plenty of riding in Switzerland (notably the Jura) and even more in Austria. Most riding schools rent out horses in winter as well as summer.

Sailing Most lakes of any size have dinghies for rent – ask at local tourist office. To rent one, you may need to be able to prove competence, *eg* with an A licence. There are restrictions on the size of boat you can take to each country. For competition sailing, contact the Austrian Sailing Association, Grosse Neugasse 8, 1040 Wien, or Swiss Yachting Union, Schwarztorstrasse 56, 3007 Bern.

Skating There are about 250 ice-rinks in Austria, rather more in Switzerland. Also skating on the frozen lakes. Skates can be rented fairly readily. No need to look for any particular place for this sport, but Davos is the skating capital.

Skiing The practice of sliding along the snow on long thin planks is big business – it brings in 40–50% of the foreign visitors to Switzerland and Austria, and the numbers are increasing every year. Austria and Switzerland have traditionally been the places for individual skiing holidays rather than for package tours, but this has changed in recent years. For example, there are about 400 small towns and villages in Switzerland with enough reliable slopes and enough ski lifts to claim to be a ski resort, but there are only about 70 of these which offer sufficient accommodation to interest tour operators who rely on mass tourism. These are the names one hears of, which then become the 'in' places, and therefore expand further, though they may be intrinsically no better or worse than anywhere else. In Austria there is a handful of 'big name' resorts, which became established because they had the best slopes when skiers were a rare breed and they have remained fairly expensive, and there are the resorts being built up by mass tourism such as the Swiss, French and Italian resorts, but there are also many hundreds of little villages, especially in Kärnten and Steiermark, which get good snow and may have a lift for the locals, but don't pretend to be resorts. The quality of the nightlife, an inescapable part of a skiing holiday, varies from resort to resort and from year to year, and you learn about this on the grapevine once you have done a bit of skiing. But for the beginner, the only economical course is to take a package tour because you can then rely on there being a ski school at your chosen resort – it's no good finding yourself an undiscovered rural idyll if you can't go out on your own. Package tours use resorts whose altitude makes it fairly reliable that there will be snow for skiing during the booked months.

A superficial rundown of ski resorts is no help at all, so for a review of resorts from the skier's, rather than the tourist's, point of view, get a publication like the *Daily Mail Skiing Guide*.

There are four varieties of the sport: downhill, cross-country, summer, après.

Downhill skiing (called 'ski' in French, 'Schi' in German) is the glamour sport, unavoidably expensive because all those hoists cost money, but it need not be so expensive as to become exclusive. It is very exhilarating and if you are in the alpine countries in winter as a tourist, you should try it. The basic equipment – skis, sticks, boots and bindings – can be rented in most towns and at the skiing centre; renting is often more expensive at the centre, but preferable because if the boots aren't quite right for you they can be changed. Clothing should be warm, waterproof and allow freedom of movement – sweater and anorak will do, there's no need to look like a skiing ad until you become more proficient. Dark glasses are needed when you get out into open country; very dark reflective glasses are for high-slope skiers in brilliant sun; goggles are for certain atmospheric conditions, not for beginners; it is common to wear no glasses on nursery slopes.

In general, costs are not quoted in this book because inflation confounds us all, but since there is no comparable skiing in English-speaking countries here are some 1982 prices as a guide.

One week's ski insurance (excludes jumping)	£10 or $25
One week's rental of boots	120–150 ÖS or 35 Sfr
One week's rental of skis and sticks	250 ÖS or 70 Sfr
Six-day ski pass for unlimited use of hoists	750–1000 ÖS or 150 Sfr
6 × 4 hours' ski school	700–750 ÖS or 50–70 Sfr

Cross-country skiing is usually called by its German name *Langlauf* (French: *ski de fonds*); you take long, long snow-walks along prepared paths (*Looipen*), or find your own way if you are skilled, and enjoy the countryside rather than the sport itself. In Langlauf, you are propelled by your arms rather than gravity. The skis are attached only to the toes of your boots. It is possible to learn enough in half an hour to go out on your own, so you are spared the cost of lessons and lifts, but it has neither the excitement nor the competitiveness of downhill ski. The annual Langlauf marathon in the Engadine attracts 10-15,000 participants, and the Jura, which is a bit short of downhill slopes, has many cross-country trails.

Children of four years old, maybe younger, can learn to ski and people of 80 or more still go skiing. For downhill ski it is almost imperative that you have accident insurance – many package tour operators make it a condition of booking.

Summer skiing is a way for the experienced to keep in practice. It's really glacier skiing and there are no long runs. A morning activity, because the snow can turn slushy by the afternoon under the bright sun. Many places round St Moritz are for summer skiing, Zermatt is famous, also Gstaad and Mürren (p. 97), Diablerets, Saas Fee and Crans in Switzerland; in Austria, Hintertux (Zillertal), Neustift (Stubaital), Sölden (Oetztal), Kaprun (near Zell am See) and Ramsau (Dachstein) are recognized centres. Some enthusiasts keep in practice in summer by skiing on grass, on giant roller skates.

The summer slopes, above the permanent snowline (2600-3000m/8450-9750ft) can be bitterly cold in winter.

Après-ski is the evening entertainment when the last hoist of the day has run. In early winter it's necessary because the days are short, but as the dedicated skier gets in more hours on the slopes, he may be satisfied with a bath to soak out the bruises, dinner and bed. But for many the après-ski matters more than the snow – music, dancing, chatting and drinking are where their money goes.

Swimming The lakes of southern Austria get warm in summer – enough to stay in the water for more than an hour if you are so inclined. Those of the Salzkammergut and some in Switzerland warm up sufficiently for an enjoyable dip, while the glacier-fed lakes are . . . glacial. On the whole, the higher the lake the colder, but Weissensee at 1000m/3250ft is warm. The lakes are not all so clean as some brochures suggest, but they are clean enough for swimming. Where there is no lake swimming (*Seebad*) there is in most places a swimming pool. Outdoor pools (*Freibad*) abound and are usually heated to 75°F/24°C; these are quite expensive for the occasional swim, but much cheaper if you use a facilities card issued by the resort. Indoor pools (*Hallenbad*) are less frequent; always at a spa.

Tennis Most resorts have a couple of public tennis courts where you can rent racquets. Bigger hotels in the country also have their own courts.

Tobogganing There are about 50 resorts in Switzerland and 40 in Austria with a toboggan run (*piste de luge*, *Rodel-* or *Schlittel-bahn*) and perhaps twice as many with a skibob run. Tobogganing is fun and fairly harmless, bobsledding is faster but need not be dangerous. Day's rent of toboggan about 5 Sfr, skibob 20 Sfr.

Walking One activity that the most car-bound culture vulture won't be able to resist in these mountains is walking. Whether you drag yourself 100 metres to a viewing point or spend all day and every day hiking, the clean air and the scenery are inexhaustible and invigorating. For any outing, do take sensible shoes – at least flat, lace-up and with soles that grip. It may seem unnecessary to say this until you see heedless tourists limping back to their coach. For rambling through forests, gorges and streams or hiking to any extent, spare yourself grazed knees and ricked ankles with hiking boots – worn with a pair of thick socks over your ordinary socks.

For more serious hiking follow the local custom and, instead of jeans, wear trousers fastened below the knee, leaving the calf bare or covered by long socks. Trousers can be bought locally as 'Knickerbocker'. Maps of recommended paths are available in most tourist offices and there are marked paths – at first you may want to wander at will, but the markings are based on experience and guide you along the most rewarding trail. Plenty of guided walking tours in groups.

About one third of the mountain resorts have a 'fitness course' (many of them instigated by the insurance company Vita – the fitter you are, the longer you live). This is a woodland course with exercise stations that you attempt or ignore as you please; it injects fun into a family outing.

ENTERTAINMENT

Vienna and Salzburg are the two centres of greatest international prestige for classical **music** – opera at the Viennese Staatsoper is probably the world's most glittering occasion, and the Salzburg music festival (last week in July to end of

August) covers the whole range of instrumental music, though Mozart lovers get more than their fair share. Both cities have music all year round, outside festival times; the Vienna Philharmonic and Vienna Symphony orchestras are world-famous and Salzburg, which is a large village by the standards of world capitals, provides a recital every day of the year except Sundays.

Other cities where there is usually some classical music are Geneva (home of l'Orchestre de la Suisse Romande), Basel (good modern), Zürich, St Gallen, Linz and Graz. Notable festivals of classical music are: May – Lausanne; June – Zürich; Aug – Luzern and Gstaad (Yehudi Menuhin); July to early Sept – Sion (Varga, violin) and Bregenz (light opera) and Mörbisch (light opera); Sept – Linz (Bruckner); Oct – Graz.

Jazz is not a very lively scene – even canned pop competes unsuccessfully with classical and folk music. Festivals in: April – Bern; June – Zürich; August – Willisau (near Luzern); throughout the summer, the top jazz festival is Montreux.

Nightclubs Geneva has *boîtes* that stay open as late as 0400, and in Vienna there are nightclubs or cabaret around the Kärntnerstrasse; otherwise 0200 is about as late as the bars stay open, even in the lively spots. At the more plush ski resorts in season there can be dancing and music till the early hours – a sort of avant-ski – and this is about the only reliable nightlife.

Casinos In Austria, there are gambling casinos in: Baden, Bregenz, Riezlern (Vorarlberg, approached only from Germany), Linz, Salzburg, Seefeld in Tirol, Velden (Kärnten) and Vienna, and in summer season and winter season in Badgastein and Kitzbühel. In Switzerland many resorts have somewhere for a small flutter, but the maximum stake is 5 Sfr. Konstanz across the road from Kreuzlingen, or Evian across the lake from Lausanne, and Bregenz, are handy for Swiss embarrassed by their riches.

Theatre There is regular English-language theatre in Vienna, occasional in other towns. All towns of any size have theatre in the local language.

Folklore The original purpose of yodelling may have been to call to the cows in the early evening, or more probably to sing out an evening prayer; it is reserved now for performances or contests at local festivals mainly in the German-speaking parts of Switzerland. The *Alp-horn* (*cor des alpes*), 3 metres long, is another beloved symbol of the Swiss past, still sounded at festivals. Wrestling Swiss-style (*lutte Suisse*, or *Schwingen*) calls for beefy men in canvas trousers, and a few bouts between local farmers enliven ceremonial turnouts; wrestling Austrian-style (*Preis-Ranggeln*) is still seen in Tirol. The Austrians more than the Swiss go in for bands (*Kapelle*), who usually wear a uniform derived from the dress of the peasant-soldiers of the heroic days of the early 19th century; every village has its band, who turn out on any excuse or none. *Schuhplattler* dancing, by thigh-slappers in Lederhosen, is occasionally performed on stage at private festivities, but has become something like flamenco in Spain – laid on by restaurants for visitors; Tirol is the region for Schuhplattler. Plain dancing tends to be disco, but if advertised as *Jägertanz* (hunters' dance), you may see the traditional dances of the region, in traditional costume.

Traditional costume (*Tracht* for women) is worn as Sunday best and for festivals and folk events in many parts of the mountains – the Valais, Appenzell and Salzkammergut, and the Lungau district of Salzburg, should be specially mentioned. To the uneducated eye, all these costumes are variations on an Austro-Swiss alpine theme – the Dirndl skirt-with-pinafore which the locals do not regard as 'costume' at all. Two highly distinctive costumes in a different style are the broad floppy hat and cape, reminiscent of the mountains of central France, which you occasionally see in the Jura mountains of Switzerland, and the subdued Hungarian costume which is quite regularly worn on Sundays in the Austrian Burgenland. All these costumes are worn proudly and unselfconsciously as part of a living tradition.

Festivals on a village scale take place at the drop of a hat in the mountains in summer. Wherever you are, the local tourist information office will have information about established fetes within easy reach; there's something going on most weekends. But the best are those you just come across – a beer tent with an oompah band or tables and benches in the street, waiting for the locals and you, a stage in a corner of the square where the *Kapelle* or the amplifier will keep the dancers going till they drop, a bunch of students doubling up as strolling players and minstrels in Geneva or Graz, Luzern or Linz – these are the real folk events.

Here's a rundown of some of the better established folk events, whether organized or unstoppable.

January Schlittedas (costumed sleigh rides) in the Engadine; Vogel Gryff (mummers' parade and inter-Basel rivalry) in Basel; balls and masked processions in Vienna.

February Carnival (Fasching) in most places, most notorious being Basel; end-of-winter celebrations in Scuol.
March Various passion plays around Easter time.
April Meetings of the Landsgemeinde, when the entire male population meets to vote on public affairs, in communes of Glarus, Appenzell, Nidwalden and Obwalden; camellia festival in Locarno; guild parades in Zürich.
May Late May or early June is when the cows are taken from their winter quarters up to the summer pastures, paraded, sometimes bedecked, through the villages and small towns of all the alpine regions. It's associated with 'cow-fighting' (to find the cattle queen) in the Valais, and the Gauderfest, with local wrestling in Mayrhofen and Zell am Ziller. Corpus Christi is celebrated in all the Catholic areas – especially colourful in Appenzell (carpet of flowers), Kippel in the Valais (uniformed procession), Hallstatt (boat procession on the lake) and Lungau.
July Rose festival at Weggis (Luzern); crossbow shooting in Bernese Oberland and Emmental; start of the music festival season in Austria.
August and September are the great months for festivals – they are all over the place, too many to list here. The shooting contests of Zürich and the dedication of the angels in Einsiedeln deserve special mention.
September and October are the months for vintage festivals in all the wine growing regions, from Lac Léman to the Neusiedlersee. Flea market at Chaux-de-Fonds.
November Commercial fairs and open-air markets, most fun is Zürich's onion market.
December St Nicholas day (6th) celebrations in many places; the escalade (11th) in Geneva; 'Silent Night' at its birthplace, Hallein, on Christmas Eve.

WHAT YOU NEED TO KNOW

Addresses of towns are represented by a four-figure post code, preceded by CH for Switzerland or A for Austria. When the code is used, there is no need to put the province or canton on the address.

Chemists (*Pharmacie, Apotheke*) open normal shopping hours (0900 or before to 1730 or later Mon to Fri and Sat mornings in Austria; 0800 to 1800 Mon to Sat in Switzerland). In towns there is a rota of chemists open after hours (*pharmacie en service, diensthabende Apotheke*) and this rota is displayed on closed chemist's shops. If in difficulty, dial for telephone information (see p. 31).

Churches There's a Protestant as well as a Catholic church in most Catholic areas of Switzerland and vice versa. In mainly Catholic Austria there's a Protestant church in all large towns and many villages. The hours of service (*Messe* for Catholics, *culte* or *Gottesdienst* for Protestants) are displayed on a joint notice board (green outline on a white cross) at the entrance to a town or village.

Cigarettes and tobacco are a state monopoly in Austria, purchased at a *Tabak-Trafik*; mainly Austrian brands, foreign brands may be available in a large hotel or restaurant. In Switzerland there's wide choice of international brands, often manufactured in Switzerland under licence, and a very wide choice of pipe tobacco, freely available. Tobacco taxes are low in both countries.

Electricity is supplied at 220 volts 50 Hz. This is suitable for British electrical appliances, but cannot be used without an adaptor on American 110 volt apparatus and cannot be used at all on American 60 Hz timed apparatus.

Health There is no free medical service in Switzerland. There is a convention on Social Security between the UK and Austria which for British citizens provides free in-patient hospital treatment after accidents and in emergencies; otherwise visitors need medical insurance to cover possible bills. Hospitals will refer you to a private doctor unless the case is an emergency. Doctors display their consulting hours (*Sprechstunden*) outside their practices, usually five hours a day, and consultations are by appointment. Dentists keep rather longer hours. In an emergency, if you can't get to a hospital, telephone a doctor on standby, the number will be found in the telephone book under *permanence chirurgicale* or *ärztliche Notfalldienst* in Switzerland and under *Ärztenotendienst* in Austria. Otherwise dial for telephone information (see p. 31).

Rabies Rabid wild animals, principally foxes, are at large. Caution is essential when approaching any wild animals, or dogs roaming free from control. There is at present no effective preventive vaccine against rabies. A bite or scratch incurred through contact with a wild or stray animal should be washed immediately with soap and water, and medical advice should be sought.

The UK totally prohibits the importation of animals (including domestic pets) except under licence. One of the conditions of the licence is that the animals are retained in approved quarantine premises for up to six months. No exemptions are made for animals that have been vaccinated against rabies. Penalties for smuggling involve imprisonment, unlimited fines and the destruction of the animal.

Any animal being imported into the US must have a valid certificate of vaccination against rabies.

For details apply to the Ministry of Agriculture (Animal Health Division), Hook Rise South, Tolworth, Surbiton, Surrey KT6 7NF.

Lost property Report the loss to your hotel – even if lost elsewhere, they may have a bright idea. Report the loss to the police, and if an insurance claim is likely, get a written note of your report. For property lost on public transport, enquire at the *bureau d'objets trouvés* or *Fundbüro* (found property office) and leave a note of where you can be contacted because lost articles often turn up many days later. The level of public honesty in these matters is high in both countries. In Switzerland a reward of 10% of the estimated value is paid to the finder.

Newspapers English-language newspapers are sold on station bookstalls in main towns and at most of the main resorts, often on the day of publication. English-language magazines available at most magazine sellers.

Opening times Museums in Austria usually open from 1000, in Switzerland from 0900 in summer, from 1000 in winter and on Suns. Mostly closed 1200 to 1400. Closing times mostly 1730, sometimes 1800. Many in Switzerland open one late night a week, many in Austria closed Mons. Museum times in Vienna are erratic.

Small shops open from 0800 or, rarely, 0900 to 1800, often with a one or two hour lunch break, though this is disappearing. Department stores, the same hours without the lunch break. Sat afternoon closing in Austria except for food shops. Half-day closing one day a week in Switzerland (often Mon mornings) varies with the town. Austrian food shops open much longer hours.

Bars and restaurants (see p. 22); banks (see p. 14); post offices (see below).

Photography All grades of film available, at prices roughly in line with those elsewhere in Europe – an Austrian luxury tax rather pushes the price up. Black and white developing overnight in most places of any size, but colour (negative and reversal) has to be sent away, so processing is best kept until you are home. People in traditional costume or engaged in photogenic rural tasks like being photographed, so don't be afraid of asking permission – even in sign language. Photography is allowed in most museums (tripod usually forbidden, flash nearly always forbidden) but usually for a fee and you may have to deposit the camera if you don't pay the fee.

Police Austrian Federal police wear a green uniform, provincial police (Gendarmerie, seen out of towns) wear a grey-blue uniform, police in Vienna and some other towns wear green jackets with black trousers; traffic police wear white hats; motorcycle police are called *weisse Mäuser* (white mice) because of the uniform. In Switzerland there are no federal police, only cantonal and city police with a different uniform for each authority, but all are instantly recognizable.

Austrian and Swiss police impose on-the-spot fines for traffic offences and jaywalking. In both countries, but especially in Switzerland, the police are basically citizens in uniform – there to prevent breaches of the law, rather than as representatives of authority – and regarded as such by the people.

To contact the police in emergency, telephone 117 in Switzerland; 133 in Austria.

Postal services A black horn on a yellow ground is the sign of the postal administration in both countries; Swiss post offices are also identified by the Swiss cross and the letters PTT. Swiss post offices open 0730 to 1200 and 1330 to 1800 Mon to Fri (in large towns at least one post office does not close for lunch) and 0730 to 1100 on Sats; there is a post office with a late-service window (open till 2200 or 2300) in at least eight large towns, and a 24-hour-a-day office in Zürich. Austrian post offices open 0800 to 1200 and 1400 to 1800 Mon to Fri; small ones stay closed on Sats and about 400 larger ones open 0800 to 1000 on Sats. There are 24-hour-a-day, 7-day-a-week post offices in Vienna, Graz, Linz, Klagenfurt, Innsbruck, Salzburg and Villach (head office, railway station or both). Telegrams can be sent from post offices. Stamps are bought at post offices (or at a vending machine outside to avoid queues), or in Austria from a *Tabak-Trafik*, in Switzerland at station vending machines or sometimes in souvenir shops.

Poste restante (*postlagernd*) should be addressed to the main post office of the town you're visiting; passport and small change needed to collect it. In Vienna there's a single office for poste restante only, at 19 Fleischmarkt.

Public holidays Fixed dates for public holidays in Austria are: 1 Jan; 6 Jan (Epiphany); 1 May (Labour Day); 15 Aug (Assumption); 26 Oct (National Holiday);

1 Nov (All Saints'); 8 Dec (Immaculate Conception); 25 Dec; 26 Dec (St Stephen's Day). Public holidays with movable dates are: Easter Monday, Ascension Day, Whit Monday, Corpus Christi (note that Good Friday is not a public holiday). Public offices, banks, all shops, some restaurants, close on public holidays. Entertainments such as concerts may be open even on Christmas Day.

In Switzerland the only fixed dates for national public holidays are 1 Jan, 25 and 26 Dec. 1 Aug is National Day, observed as a public holiday in many cantons, but in others a partial holiday when businesses close for the afternoon. Movable dates of public holidays are: Good Friday (except in Valais), Easter Monday, Ascension Day and Whit Monday. Each canton has other days observed as a cantonal holiday, most common being Labour Day (1 May), and many local (saint's) holidays in Catholic parts. On public holidays, banks and public offices are closed, shops close early, restaurants are open late.

Pressure Pressure gauges for car tyres are calibrated either in kg/cm² or in dynes, which are so close to kg/cm² as to make no difference. 1 kg/cm² = 14.7 psi; 22 psi = 1.48 kg/cm²; 24 psi = 1.69 kg/cm²; 28 psi = 1.97 kg/cm²; 30 psi = 2.11 kg/cm².

Radio Swiss channel 1 has a daily broadcast in English at 1900, taken from BBC news. Austrian First Program transmits news in English at 0805, and Blue Danube radio (102 MHz) has a mixed output in English from 0700–0900, 1200–1400 and 1800–1930.

Shopping Top prices for top quality is the rule in Swiss shops – most products which a small country might be expected to import are made in Switzerland. Typical Swiss products are watches, cheese, chocolate, cuckoo clocks, wood carvings, *caquelon* (earthenware pot for making cheese fondue); for the most part these products are more expensive in Switzerland than when exported. There are also superb fabrics with a wider range of choice (*eg* Zürich, St Gallen) than when exported and maybe an edge on prices; also embroidery. Souvenir shops offer tourist bait such as cowbells, musical boxes, alpine scenes; well made and high priced. Best to warn the folks back home not to expect souvenirs from Switzerland.

The situation is similar in Austria, but Austrian-style clothing should be mentioned – Dirndl skirts and the Loden jackets (grey wool piped with green) most popular in Styria. Vienna can be good for antiques if your taste is for Biedermeier or High Imperial, and there's a good choice, but no bargains, of Augarten porcelain, petit point, leather goods and glassware.

Haggling is very rare – the marked price is the price. In the flea markets you can try a lower offer, or turn away with a disappointed *Zu teuer* (too dear).

Austrian Value Added Tax (18%, on some goods 30%, of the pre-tax price) can be recovered on purchases over 1000 ÖS on taking them out of the country, if the shop operates the rebate scheme and you fill in the form at the time of purchase.

Telephones (see below)

Time Both countries are in the Central European Time zone, one hour ahead of G.M.T. Both adopt summer time at *about* the same time as EEC countries, so both are one hour ahead of British time nearly all year. Their time is six hours ahead of New York, nine hours ahead of the West Coast apart from possible overlap of summer time. Adjust your watch if flying in; if arriving by road or rail your watch should already have been adjusted. The sun rises and sets in Geneva 41 minutes later than in Vienna.

Tipping All restaurant and hotel bills in Switzerland must include a service charge (15%) and this really is intended to cut out tipping; don't be afraid to leave nothing or a few small coins. Taxi fares sometimes include the service charge and then there must be a notice in the taxi to this effect and no tip, or the odd franc, is in order. If service is not included, tip taxis 10%.

Austria is trying to cut out tipping, but the heart tends to rule the head. Don't make their job harder by overtipping. Service is usually on the bill, and then around 5% extra is common; where service is not included, *eg* taxis, give 10% or, only for real service, 15%. For porters, 1 Sfr or 5 ÖS, and for washroom attendants, half that.

Toilets Public toilets are found where you'd expect them, in railway stations, restaurants, stores, also in the underground passages that are beginning to connect squares in the main towns. In case of need, you can use the facilities of bar or café without feeling obliged to buy a drink. For washing facilities there's usually a set fee.

Water Tap water throughout Austria and Switzerland is clean and safe for drinking except in rare cases indicated by *eau non potable* or *nicht Trinkwasser*.

Telephones The telephone system of Switzerland is completely on IDD (that is, you can direct-dial anywhere in the world from a private phone) and that of Austria almost completely. Telephone booths in the street can in most cases be used only for internal, not for international, calls, and so it is best to make these from a post office – ask the counter clerk which booth to use, dial your call and then pay the clerk. Calls made from a hotel are liable to attract a hefty surcharge. Telegrams are best sent from a post office, but can be telephoned if you can manage the local language (about half the telephone operators speak some English).

Swiss telephone directories have full instructions in the official languages and then abbreviated instructions in English (usually around page 9). Austrian directories have a summary of emergency numbers on the front cover but in German only. For information generally, including directory, medical, and emergencies, dial 111 in Switzerland, 16 in Austria. If the operator doesn't speak English, she will find a colleague who does.

Dial for	Switzerland	Austria
International directory enquiries	191	08
Operator for foreign calls	114	09
Telegrams	110	10
Ambulance	144	144
Police	117	133
Fire	118	122
Motoring assistance (*Pannendienst*)	140	*
Tourist information, snow report	120	192
Road conditions (*Strassenzustandbericht*)	162	194 or 19
Time	161	15

* For motoring assistance in Austria see the number given under *Pannendienst* on the front page of the province telephone directory.
External dialling codes: Australia 0061; Canada 001; Ireland 00353; New Zealand 0064; S. Africa 0027; United Kingdom 0044; United States 001. British numbers drop the '0' which precedes the area code (01, 021, 031). The four tones you hear on the phone are: a long steady tone – ready for dialling; long tones interrupted by a short pause – number is ringing; rapid short tones – number is engaged; three tones in rising sequence – fault, dial information.

USEFUL ADDRESSES

For further information on either country, your best source is the Austrian or Swiss National Tourist Office, here abbreviated to ANTO or SNTO. Some of their addresses (tel. nos in brackets) are:

SNTO
Canada: Commerce Court West, Suite 2015 (PO Box 215), Toronto, Ont. M5L 1EB ((0416)868 0584). **UK:** Swiss Centre, 1 New Coventry St, London W1V 3HG (01-734 1921). **USA:** The Swiss Center, 608 Fifth Ave, New York, NY 10020 ((212)757 5944); 250 Stockton St, San Francisco, CA 94108 ((415)362 2260)

ANTO
Canada: 2 Bloor St East, Suite 3330, Toronto, Ont. M4W 1A8 (967 3381); 1010 ouest rue Sherbrooke, Suite 1410, Montreal, Quebec H3A 2R7 (849 3709); Suite 1220–1223, Vancouver Block, 736 Granville St, Vancouver, Brit. Columbia V6Z 1J2 (683 5808). **UK:** 30 St George St, London W1R 9FA (01-629 0461). **USA:** 545 Fifth Ave, New York, NY 10017 ((212)697 0651); 200 East Randolph Drive, Suite 5130, Chicago, IL 60601 ((312)861 0100); 3440 Wilshire Blvd, Suite 906, Los Angeles, CA 90010 ((213)380 3309); 1007 N.W. 24th Ave, Portland, OR 97210 ((503)224 6000). **S.Africa:** The Trust Bank Centre, 30th floor, Eloff St, PO Box 7999, Jo'burg 2000 (21 11 37). **Australia:** 19th floor, 1 York St, Sydney NSW 2000 (27 85 81). **Ireland:** 4 Ardoyne House, Pembroke Park, Ballsbridge, Dublin 4 (68 33 21).

Within the countries, there is a tourist office of some sort in nearly every resort, but these deal mainly with their own resort. For general information consult a *regional* tourist office in Switzerland or a *provincial* tourist office in Austria. The eleven Swiss tourist regions are: **Grisons** Hartbertstrasse 9, 7001 Chur. Tel: (081) 22 13 60. **Eastern Switzerland** (Cantons of Appenzell Ausser-Rhoden, Appenzell Inner-Rhoden, Glarus, St Gallen, Schaffhausen, Thurgau and Principality of Liechtenstein) Bahnhofplatz 1a, Postfach 475, 9001 St Gallen. Tel: (071) 22 62 62. (Leichtenstein) Postfach 139, 9490 Vaduz. Tel: (075) 6 61 11. **Zürich** Bahnhofbruecke 1, 8023 Zürich. Tel: (01) 211 12 56; (information and travel service) Bahnhofplatz 15, 8001 Zürich. Tel: 211 40 00. **Central Switzerland** (Cantons of

Luzern, Nidwalden, Obwalden, Schwyz, Uri, Zug) Pilatusstrasse 14, Postfach 191, 6002 Luzern. Tel: (041) 23 70 45. **North-Western Switzerland** (Cantons of Aargau, Basel-Landschaft, Basel-Stadt, Solothurn) Blumenrain 2, 4001 Basel. Tel: (061) 25 38 11. **Bernese Oberland** Jungfraustrasse 38, 3800 Interlaken. Tel: (036) 22 26 21. **Bernese Mittelland** c/o Offizielles Verkehrsbuero der Stadt Bern, Bahnhofplatz 10, Postfach 2700, 3001 Bern. Tel: (031) 22 12 12. **Fribourg/ Neuchâtel/Jura** (Cantons of Fribourg, Neuchâtel, Jura plus Bernese Jura) (Fribourg) 8 Route Neuve, Case postale 901, 1701 Fribourg. Tel: (037) 23 33 63. (Neuchâtel) 9 Rue du Trésor, Case postale 612, 2001 Neuchâtel. Tel: (038) 25 17 89. (Jura) 16 Rue de l'Hôtel-de-Ville, Case postale 338, 2740 Moutier. Tel: (032) 93 18 24. (Bernese Jura) 26 Avenue de la Poste, 2740 Moutier. **Lake Geneva Region** (Cantons of Geneva, Vaud) (Geneva) 1 Rue de la Tour de l'Ile, Case postale 440, 1211 Genève; (for personal callers) 2 Rue des Moulins, En l'Ile. Tel: (022) 28 72 33. (Vaud) 10 Avenue de la Gare, 1002 Lausanne. Tel: (021) 22 77 82. **Valais** 15 Rue de Lausanne, 1951 Sion. Tel: (027) 22 31 61. **Ticino** Piazza Nosetto, Casella postale 643, 6501 Bellinzona. Tel: (092) 25 70 56.

The nine Austrian provincial tourist offices are: **Vienna** Kinderspitalgasse 5, A-1095 Vienna IX. **Burgenland** Schloss Esterházy, A-7000 Eisenstadt. **Kärnten** Kaufmanngasse 18, A-9010 Klagenfurt. **Niederösterreich** Strauchgasse 1, A-1014 Vienna I. **Oberösterreich** Johann-Konrad-Vogel-Strasse 2, A-4010 Linz. **Salzburg Prov.** Mozartplatz 1, A-5010 Salzburg. **Steiermark** Herrengasse 16, Landhaus, A-8010 Graz. **Tirol** Bozner Platz 6, A-6010 Innsbruck. **Vorarlberg** Römerstrasse 7/I, A-6901 Bregenz.

Austria also has regional tourist offices, but these just cover a small group of neighbouring resorts.

Consulates(c) and Embassies(e)

Australia: Vienna 528586 (e); Bern 450153 (e); Geneva 246200 (c). **Canada:** Vienna 633691 (e); Bern 03144 (e). **N.Zealand:** Vienna 264481 (e); Geneva 349950 (e). **S.Africa:** Vienna 630656 (e); Innsbruck 22379 (c); Bern 442011 (e). **USA:** Vienna 315511 (e); Bern 437011 (e); Geneva 346031 (branch office); Zurich 552566 (c). **UK:** Vienna 731975 (e); Innsbruck 05222-3-20 (c); Bern 44502 (e); Geneva 343800 (main c); Montreux 541207 (vice c); Lugano 545444 (c general); Zurich 471520 (c). **Ireland:** Vienna 754246 (e).

Motoring Organizations

Touring Club Suisse, 9 Rue Pierre-Fatio, 1211 Genève 3 ((022)36 60 00), affiliated to AA.
Automobil-Club der Schweiz, Wasserwerkgasse 39, 3000 Bern 13 ((031)22 47 22), affiliated to RAC.
ÖAMTC: A-1010 Vienna, Schubertring 3 ((222)72 99-0).
ARBÖ: A-1150 Vienna, Mariahilfer Strasse 180 ((222)8535 35).

LANGUAGE

In Austria, about 4% of the people, concentrated in the southeast near the Yugoslav border, speak Slovene, otherwise everyone speaks the sole official language, German. There is no substantial difference in writing between the German of Germany and the German of Austria, but considerable difference in pronunciation. The Viennese accent has a highly individual and easily recognizable soft lilt, which can also be heard in Niederösterreich; elsewhere city dwellers have a sort of mild Bavarian accent. The rural accent in mountain districts is often very strong, with long vowels, softened consonants and the ends of the words left off, so that even a city Austrian may have to strain to understand. Speakers with a strong rural accent usually try to speak *Hochdeutsch* when talking to a foreigner and the result is what a North German would consider a typically Austrian accent.

Hochdeutsch is the German equivalent of BBC-English, without the class implication. About 65% of Swiss people speak German, mainly in the north, east and centre, but it is certainly not Hochdeutsch. When written down, as in a newspaper, it looks exactly like the German of Germany, but an educated German-speaking city Swiss can sound broader than the most rural Austrian. And the strong accent of a remote village can be unintelligible even to the next village let alone to the city dweller, still less to the tourist who learned a bit of German at school. These local varieties of German are called *Schwyzerdutsch* and a toned-down version of Schwyzerdutsch is used on radio and television. In talking to a foreigner, *ie* anyone not from his own canton, a German-speaking Swiss will attempt Hochdeutsch.

French is spoken by about 18% of Swiss people, in the Western Region described in this book, and in the Valais

Languages

westwards as far as Sion. Canton Fribourg is partly German, and Biel/Bienne is totally bilingual. The language is the same as in France with a few idiosyncrasies. The accent is perfectly clear, good French: indeed the purest French of any French-speaking territory is spoken in Neuchâtel. (The same claim is made for the French of Geneva and the French of Lausanne.) There is a rural accent in the Valais which is based on Provençal, not on Northern French.

Throughout canton Ticino (apart from the influx of rich, sun-loving, retired Germans) and in parts of Graubünden, Italian is the local language, spoken by about 12% of Swiss people. It is standard Italian, with an everyday Lombard accent.

German, French and Italian, then, are the three official languages. They have, in theory, equal status; all three are used in publications of the Federal Government and federal officials must be fluent in any two. In practice, German tends to come first and Italian is always third. But the German-speaking Swiss are careful not to impose on the minority languages, while the minorities are at pains to feel Swiss, and not culturally part of France or Italy. This need to compromise, to avoid any obvious top dog, goes deep into the Swiss nature. It appears, of course, in religious, social and economic compromise as well.

There are two acknowledged principles governing the use of language, and the subtle way they are applied is a lesson in how to get on peaceably. The personal principle is that each person has the right to use and be addressed in his mother tongue (after a few sentences it is easy enough for the Swiss to agree tacitly which language is easiest to use). The territorial principle is that each commune, and thus, in most cases, each canton, can decide which language will be used, *eg* in the schools. Official notices of the canton and the commune need be in only one language. The territorial principle is the more important and on trains and buses you will hear the guard or driver switch from one language to another in mid sentence, as the vehicle passes the unmarked language boundary.

In addition to the three official languages there is a fourth national language (but 'national' is a meaningless flattery). This is Romansch, a relic of Latin, spoken only in canton Graubünden by a mere 60,000 people as explained on p. 82. Romansch used to be spoken in the western Tirol and there are still a few thousand people there who understand it. But the unofficial fourth language of Switzerland and the second language of Austria is English.

Throughout the tourist track in both countries, and in hotels, restaurants, banks, big shops, airports, ski schools, *etc*, English is spoken easily, almost automatically, and if that is your only language you will have no need of any other. Many Austrians at all levels – in garages, small shops, remote Gasthöfe, the man in the street – have some knowledge of English and are keen to practise it unless you show a determined competence in German. The Swiss speak English rather less well – but after all, they have their own three languages to master first.

In Switzerland timetables, notices in rental centres, menus and opening times are often given in German, French, Italian, English – or sometimes just in two languages, one of which is English. In Austria, German and English are very often used side by side on the tourist beat. But you will enjoy your stay much more if you make some attempt at the local language. Some words from a good phrase book, such as Collins French, German and Italian phrase books, serve to break the ice and foster friendly relations.

Walking in the mountains, people always pass the time of day as they meet. In Austria the greeting is *Grüss Gott*, in German-speaking Switzerland it is *Grüssli* or *Grützli*.

WIEN

(pop. 1,650,00) Former capital of the Habsburg Empire, and, since 1919, capital of the Austrian Republic.

Wien, Wien, nur du allein
Wirst stets die Stadt meiner Träume sein
'Wien, Wien, just you alone will ever be my dreamland home.' To get the full effect of this catchy, sentimental waltz-song, usually known in English as 'Come, come, Vienna mine', you have to call the city by its German name, pronounced 'Veen' with a single pure 'ee' sound drawn out until the singer's lungs are emptied and the fiddler runs out of bow. You will still hear this song, and many like it, played in the wine taverns north of the city and in the parks, for it is a faithful expression of what the Viennese feel – fond of Gay Vienna, in love with its monuments, praising it half-sincerely for virtues it may once have had. The Viennese have mixed, uncertain feelings, that come out as irony, friendliness, bravery without heroics, and courtesy with style.

When was 'Gay Vienna'? It seems always to have been a legend, alive in the preceding generation; if Gay Vienna is still there, only the next generation will realize it. The legend dates back at least to the Congress of Vienna (1815) when the most glittering array of crowned heads, courtiers and diplomats met to settle the future of Europe once and for all, following the downfall of Napoleon. Their plans were cut short by his return, but their enduring legacy was the recollection of the brilliant balls and the round of pleasure so essential to diplomacy. '*Le congrès dance, mais il n'avance pas,*' was one of the disaffected quotes of the time.

After the Congress, Vienna settled down to over thirty years of peace; they call this the Backhendel time; the age of roast chicken. It was then that the elder Strauss created the Viennese waltz which set the world alight and established Vienna's reputation as the international centre of gaiety. Unassuming mediocrity and small-scale domestic virtue were supreme in the Biedermeier period, noted for its solid and conventional furniture, much admired by the petty bourgeoisie. A vital element of the time was the devotion of the Viennese to their Emperor who ensured this comfortable, frivolous existence. This era is called the *Vormärz*, the pre-March; in March 1848 the Viennese showed their 'loyalty'by chasing the Emperor out and manning the barricades. From then on most thinking people knew that, though the Emperor would return, the monarchy was dying.

From the late 1850s there was an age of renewed prosperity, when freedom and liberalism advanced, and constitutions were freely granted, each with a clause permitting the Emperor to govern by decree. This is called the Ringstrasse age to mark the building of the Ring (p. 35) Vienna's most inescapable monument to the Habsburgs. Vienna was the centre of a phoney Empire which could not survive, and as the gaiety accelerated, people retreated further into irony and cynicism.

At the turn of the century the smell of decay was unmistakable over the remorseless fun, elegant shopping, irresistible music, and the fermenting intellectual life based on the mixture of all peoples of the Empire, crucially spiced with Jews. But it took the defeat of the First World War to bring an invitation to the last Emperor to leave: with ironic courtesy 'Your taxi is waiting, Herr Habsburg'.

Between the wars, looking back to the departed Habsburgs or looking forward to the coming of Hitler, was clearly not the time of Gay Vienna, nor were the years after the Second World War when the Allies occupied part of the city and the jeeps of four powers patrolled the Inner City. But in 1955 the Allies withdrew and life began again with a gala performance of *Fidelio* at the restored Opera House.

The physical weight of the past, in the form of the palaces and court of the Inner City, the emptily imposing Ring around it, and the functionless aristocratic palaces further afield, may oppress the visitor but it does not oppress the Viennese, it is an essential side of their life. Vienna is not just a physical museum, but a museum of attitudes.

On many monuments from the past you will see the letters KK which need to be explained. In 1867, the Habsburg lands were divided in two – the Empire of Austria and the Kingdom of Hungary, with separate governments but with a single army and a single King-Emperor who was the State. Anything belonging to the State was called Kaiserlich (Imperial)-und-Königlich (Royal), abbreviated to KuK. As it became apparent that the dual monarchy was doomed to break up, the title developed into a mystic symbol of unity, the single word Kaiserlich-Königlich, abbreviated to KK. These letters are used as a public joke today, being the initials of the president (Kirchschläger) and the chancellor (Kreisky), and they are also used as a private irreverence by street traders who label themselves KK – by royal appointment – equivalent to the British H.M.

The uncertainty of Vienna's position may explain the mixture of frivolity and cynicism that is one side of the Viennese character. Vienna was intended by nature not as a capital but as a frontier town, a defence of Western Europe against the hordes from the east: Avars, Hungarians, Turks or Russians. But in time of peace a frontier town is a bridge, and that is how Vienna functions today – a bridge between east and west. The constant contact with different peoples and their uncertain history created the Viennese who are open, sentimental, realistic, courteous, fond of display and dressing up (the Vienna opera is the last in the world where full evening dress is obligatory on some nights) and, above all, *gemütlich*. Perhaps Vienna is as gay as ever it was.

The Ring

Most imposing edifice of the grand museum that is Vienna is not a building at all, but a street – the Ringstrasse. It runs as a broad belt round the Inner City, along the line of the old city walls, virtually cutting off the centre from the rest of Vienna and imposing itself on the whole feel of the city. The monumental buildings along the Ring may be no delight to look at in themselves (though Hitler as a young man spent hours enchanted by their grandeur), but their effect is to draw the eye to the street itself; the street dominates the buildings. The Ring is 57m/185ft wide, marked out by four lines of lime trees, and a superb sight. Traffic is heavy, of course, but traffic lights allow pedestrians to get across. (No jaywalking – if you are not knocked down you will be fined.)

When Haussmann built the great boulevards of Paris, his purpose was not just to create grand streets, but to provide a cordon by which the city could be controlled when revolution came. Vienna's Ring was created for just the opposite reason. All the way round the city walls there used to be a wide open space, the *glacis*, originally created as a defence against invaders, to be used by the army to seal off the Inner City. But in the 1860s Emperor Franz-Josef ordered the space to be turned over to civilian use and to connect city to suburbs; the building spree that followed was almost the last surge of life in his dying empire. Military insulation was lost, though, socially, the Ring still isolates the city.

The Ring is over 4km/2¼mi long, from the upper end of the quay along the Danube Canal to the lower, but you can get the general effect by taking a tram.

The Ring begins by the Danube Canal with a stretch called the Schottenring, named after the Scoti (Irish Benedictines) who built the nearby Scots' Church (p. 40) in the Inner City. The first monumental building on the Ring was the **Votive Church**, built, in French Gothic style, as a garrison church for Vienna. Here are the sad memorials to the dead of the KK regiments of the First World War, to the SS regiments that died for Führer and Vaterland in the Second World War, and to the Austrian resisters who fell to the Gestapo.

Then comes a stretch called first Karl Lueger Ring and then Karl Renner Ring. An ironic juxtaposition – Lueger was the anti-Semite who became mayor of Vienna in 1895, though many of his best friends were Jews, while Renner was the socialist who became president of the first Austrian republic on the collapse of the Habsburg empire, and president again of the second after the collapse of Hitler's. The four buildings here are: the **University**, which looked to its sponsors like a Renaissance centre of culture but to us looks like a Victorian pile; the **Court Theatre** (Burgtheater), another Renaissance effort, which offers very good theatre in its magnificent interior; the **Town Hall** (Rathaus), opposite the Theatre, which is in the Gothic style (west of the Town Hall is the Austrian Folk Museum); and most spectacular, the **Parliament** building (Reichsrat). This has the façade of a Greek temple, though on a monumental scale; in front of it is a statue of Athena the Greek goddess of wisdom. The statue is very informative: first, why is she outside? Because, according to an old Viennese joke, you won't find wisdom inside; and second, why not a statue of some great parliamentarian? Because there were no great or small parliamentarians in Imperial Austria.

These four buildings are grouped round the Rathausplatz, where there are good open-air concerts in summer. There are also concerts in the Rathaus courtyard and a Christmas fair and markets.

Next stretch is the Court Ring (Burgring) which faces the Court buildings in the Inner City. This central stretch is the only part of the Ring where the buildings relate to each other and not to the street. Facing each other across the Mariatheresienplatz are the **Museum of Natural History** and the **Museum of Fine Art** (Kunsthistorisches Museum) which has a superb collection of all the established European masters; this rivals the Louvre and the Hermitage, and is a must-see for culture 'sparrows'. On the inner side of the Ring, opposite these museums, is the **New Court Castle** (Neue Burg), one half of a planned extension of the Court which was intended to link up with the museums in the world's most grandiose town plan-

36 Wien

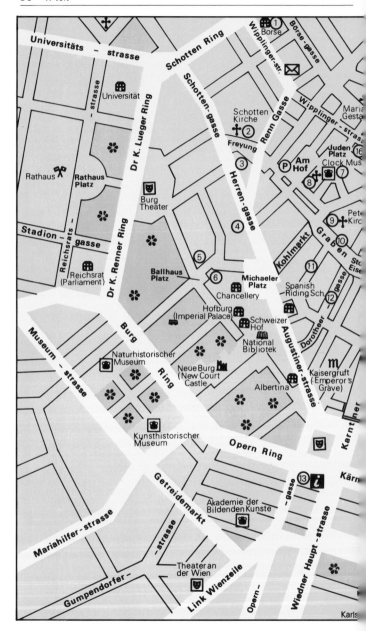

Wien 37

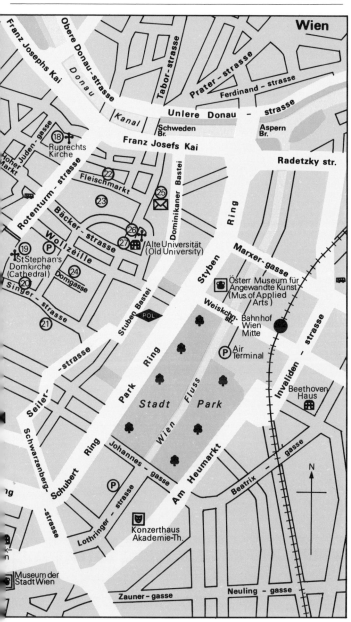

ning. They ran out of funds before attempting the matching half, and instead there stands the **Heroes Park**, where the horse cabs (*Fiaker*) wait to take tourists round the Inner City. The cabbies wear bowler hats, a charming smile, and a Viennese accent thick enough to camouflage their flow of falsehoods. There are some interesting lesser **museums** in the Neue Burg – armour, ethnography, musical instruments, including Beethoven's piano, and especially the collection of antiquities, with the reliefs recovered by Austrian archaeologists from Ephesus (at Ephesus the prime exhibit is the library restored by the Austrians).

After the grandeur of the Court section comes the Opera Ring (Opernring); it is lined with private houses built with the rest of the Ring development in a lavish style to reflect the importance of the occupants. These houses are called *Mietpalast* (rent-palaces), and the term tells you a lot about Vienna past and present. Previously, a *Palais* (palace) was a private house built for occupation by a single owner – a nobleman flaunting his station in the building; lesser beings had to live in a tenement – a *Miethaus* (rent-house). It was the building boom of the Ring that produced the contradiction in terms, the rent-palace – built from the start as a huge multi-tenanted block, but with the appearance of a nobleman's town house adapted to the liberal era. Further away, where the rich didn't see them at the time and where you won't find them today, was the *Mietkaserne* (rent-barracks), a cramped block of flats with facilities from the Middle Ages. The Opera Ring ends in the **State Opera House**; from the outside it looks yet another imposing Renaissance-style block, and it has an interior of sumptuous magnificence. The Opera was the first building restored after the destruction of World War II. A night at the opera is one of the highlights of a visit to Vienna, but you need to book tickets well in advance; on gala nights, black tie (tuxedo) is obligatory and white tie (full evening dress) preferred.

The Opera House stands on the corner of Kärntnerstrasse, the principal road south out of the Inner City, and after that the street is called Kärntnerring. From here on the Ring is less imposingly developed – private money ran out after the crash of 1873. Most of it is the Parkring, alongside the town park where there is a little lake, an open stretch of the river Wien, and **memorials** to Schubert, Bruckner, and Johann Strauss who created the waltzes that you still hear in the parks of Vienna. The final stretch is the Stubenring, with the **Museum of Applied Arts** (carpets, enamel, old Vienna porcelain) on one side, next to the former War Ministry, and, on the other, an interesting sign of things to come: the Savings Bank built in 1906, which broke away completely from the monumental, backward-looking styles of the Ringstrasse – it looks more like a 1920s building.

The Ring ends, as it begins, at the Danube Canal. The Danube used to be a number of different streams at this point, regularly flooding the marshes in between, and Vienna developed on the highest ground by one small arm of the river. The streams were straightened up long ago into one main course, the muddy grey Danube (*Donau*), and just one arm is left to flow past the Inner City under the name Danube Canal (*Donaukanal*). The road alongside the river, lined with cafés but screened from a view of the river by an embankment, is the Franz-Josefs-Kai.

Opera House

The Inner City

Old Vienna is the area enclosed by the Ring and the Danube Canal. Here are concentrated most of the sights from the past and much of the life of the present. It's a mixture of little back alleys, relics of medieval and even Roman Vienna, proud Baroque churches, prouder palaces of a vanished nobility, pedestrian precincts for elegant leisure, busy commercial roads – all jumbled up and overlooked by the spire of St Stephans Cathedral (Stephansdom) – right in the middle and dependent on the vast complex of the Imperial Palace.

Parking in the Inner City is, in the classic Viennese phrase, hopeless but not serious. If you're prepared to push, and disregard the regulations, you will find somewhere to park after an hour or so. There's an underground car park at the Opera, in the middle of the Ring, and a surface car park halfway along Franz-Josefs-Kai by the canal. But it's better to take the underground from outside and arrive at the very centre – Stephansplatz. (There are no buses in the Inner City.) At the underground station there are the remains of the Roman church, uncovered during building of the underground railway, which was the site of St Stephans Cathedral; these can be seen from a viewing gallery at the station booking hall.

This section should be read in conjunction with the map on p. 36–7.

St Stephans (19) was built around 1270 by Ottokar, the Bohemian king who seized Austria for a short while until the Habsburgs came looking for a building site. (Ottokar's ancestor was Good King Wenceslas, who looked out on the feast of St Stephan.) The cathedral was badly damaged by bombs and artillery at the end of World War II, but completely rebuilt with voluntary contributions from every part of Austria. The steeple rises 137m/445ft above the little Stephansplatz, and is a landmark from the heights surrounding Vienna. The interior is not all that interesting, and comprehensive guides are available at the entrance. The best parts are the catacombs, lined with urns containing the organs of Austrian emperors, and the towers – the main tower entered from the south side, and the north tower where hangs the largest bell in Austria, the *Pummerin* (boomer).

To help you explore the rest of the Inner City here's a list of the streets in which there's some notable building (the Imperial Palace is dealt with separately on p. 44); they are in alphabetical order. There are many suggested tours of this district, eg in the little book *Introducing Vienna* which you can buy from the tourist office, but the back streets are such a maze there's no logical order in which to proceed – pick out what interests you, and plan your route by map.

Am Hof is the largest square of the Inner City, and was the market place, execution square, jousting stadium, and palace courtyard of the Babenberg Dukes, Austria's favourite rulers 100 years before the Habsburgs. It was here in 1806, in the church of the Nine Choirs of Angels (8), that Emperor Franz II renounced the throne of the Holy Roman Empire, and became Emperor Franz I of the Austrian Empire. And here, in 1848, the Minister of War was hanged from a lamp-post by Viennese revolutionaries. Next to the church is the Clock Museum (7), including Rutschmann's great astronomical clock of 1769, and opposite is the former Armoury, now the main fire station.

Bäckerstrasse is lined with houses of the 16th and 17th centuries – see the Renaissance courtyard at No. 7; now mainly occupied by antique galleries and art dealers, it houses also the Old University, founded in 1365, cleared of students after the revolution of 1848, and now the Academy of Sciences (27); just off it is the Jesuit Church of the Assumption (26), rich Baroque dating from 1627.

Ballhausplatz has one end of the Palace (6) on one side, and the Federal Chancellery (5) on the other; Metternich dictated policy from here, and 'Ballhaus' meant one of the world centres of power, equivalent to the Kremlin or White House today.

Michaelerplatz from Hofburg

Domgasse In this little cobbled alley, at the back of St Stephans, lived Mozart, at No. 5, now labelled the Mozart Denkmal (24). Here he received Haydn, gave lessons to Beethoven, and wrote *The Marriage of Figaro* – the house is also called Figarohaus.

Dorotheergasse is a pedestrian precinct which houses the Dorotheum (12), once the State pawnshop, where the best antique auctions are held.

Fleischmarkt is a wandering narrow street, the former meat market. Where it meets Griechengasse is the Griechenbeisl (22) (Greeks' Tavern), almost the last of the cafés where musical customers such as Beethoven, Schubert and Strauss used to gather.

Freyung is a square with the little palace of the Harrach's (3) on one side (Haydn's mother was cook here), and the Church of the Scoti (2) on the other – the Scoti were Irish monks who built the church here in the 12th century.

The Graben

Graben is a pedestrian precinct, a centre of café life, lined with expensive shops. The column in the middle (10) is a memorial to salvation from the plague, erected in 1682. St Peter's church (9) is very Italian, intimate Baroque. Heiligenkreuzerhof (23) houses the 17th-century convent of the Cistercians, peaceful and popular with writers and artists.

Herrengasse, connecting the Freyung to the Palace, is only a narrow street but the most aristocratic in the city. It is lined with column-fronted town houses (palaces) of the old aristocracy, and is a main traffic artery; the palaces are now government Ministries and there is also the Museum of Lower Austria (4) with good representation of old Vienna.

Hoher Markt (17) is the site of the Roman forum, dating from the time when this was the city of *Vindobona*, where Marcus Aurelius died in AD 180. Bombing in 1945 exposed two legionaries' houses here (at No. 3), and the Roman excavations beneath Hoher Markt are now preserved.

Judengasse (Jews' Alley) runs between the Hoher Markt and St Ruprecht's church (00), the oldest in Vienna, built around the 11th century, pure Romanesque. It's still a good street for second-hand clothes, in the unfashionable bit of the Inner City between the cathedral and the canal.

Judenplatz (Jews' Square) was once the centre of the Jewish quarter, and is still the centre of the garment-making industry.

Kärntnerstrasse

Kärntnerstrasse is the most fashionable street in Vienna and has some of the most elegant shops (though the main shopping street, with the big stores, is outside the Inner City, the Mariahilferstrasse). Off this street are little alleys with nightclubs and cabarets. This was once the main road south from St Stephans, crossing the city walls where the Opera House now stands. To Prince Metternich, the great statesman, those walls marked the end of civilized life. 'The Balkans', he said, 'begin at the Kärntnerstrasse' – recognizing the position of Vienna as a frontier city. Today it is a pedestrian precinct, with pavement cafés, painted houses, painted porcelain and painted ladies.

Demel's Cafe, Kohlmarkt

Kohlmarkt, a pedestrian precinct between the Palace and the Graben, houses the famous Demel's café, one of the greatest coffee houses of Vienna, which fought with Sacher's (the other) for the right to make *Sachertorte*. Another street of shops and leisure.

Wien 41

Rotenturmstrasse is a main traffic highway, marking the edge of the old Roman city. Buildings are heavy, grey 19th-century, becoming less imposing as they slope down towards the canal. In the middle is Lugeck Square, a haunt of the café set, where stands the statue of Gutenberg, inventor of printing.

Singerstrasse houses the 14th-century Deutschordenskirche (20) (Church of the Teutonic Order) which, amid all the Baroque and Imperial splendour of Viennese churches, is austerely simple, white-painted, reverent. The Order was active around the Baltic, colonizing East Prussia and Lithuania and spreading German culture as far as the river Memel, but its Master was always the Holy Roman Emperor of the German Nation, who was usually a Habsburg. The treasure of the Order is on view in a suite above the church. In Singerstrasse, too, is the Fähnrichshof (21), a charming little courtyard which was completely renovated after bombing and is now a miniature artists' quarter, with studios, galleries, and apartments grouped round the gardens.

Stock in Eisen Platz (wood-in-iron square) is a small extension of St Stephansplatz, surrounded by modern shops. It preserves the trunk of a tree into which medieval journeymen used to drive a nail to ensure good luck on returning to Vienna.

Wipplingerstrasse is another main road into the centre, from the Börse (1) (Stock Exchange). It crosses as a viaduct over the Tiefer Graben, and just after the viaduct you can see the church of Maria am Gestade (14) (Mary by the Banks). It stands on a terrace which used to overlook the Danube before canalization, and has the most delicate stonework of any church in Vienna and has a famous, 15th-century painted altarpiece; it was built in Flamboyant Gothic at the end of the 13th century. Further along Wipplingerstrasse is the Old Town Hall (15) with the Bohemian Court Chancellery (16) facing it, dating respectively from the 14th century (lovely fountain in the courtyard), and early 18th century.

Wollzeile, once the base of medieval wool merchants is now a busy commercial street.

The main post office (25) is in Dominikanerstrasse, the Lost Property Office (11) is in Braunerstrasse, facing the Plague column in Graben, the Tourist Office (13) is in Opernpassage, the vast pedestrian subway that connects the Opera Ring with Karlsplatz underground station.

Peterskirche and Stephans Dom: top; from the Am Hof to the Danube Tower

42 Wien

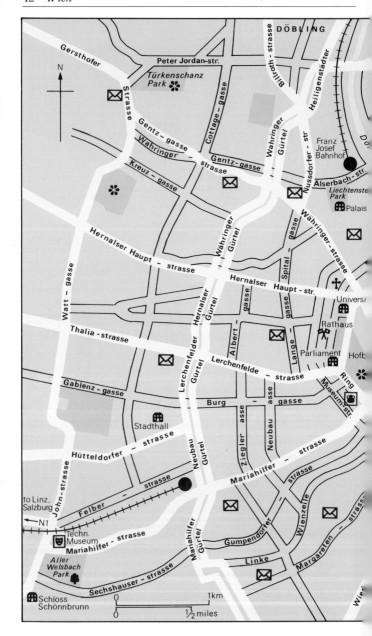

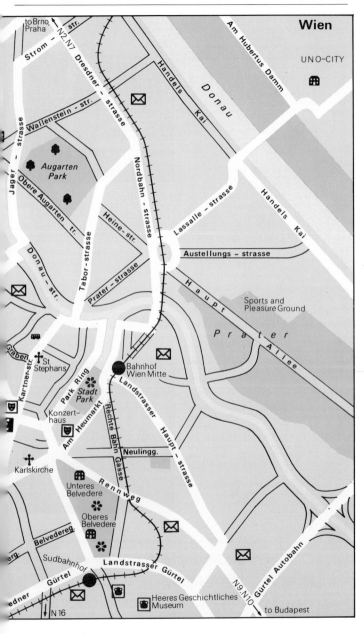

The **Imperial Palace** (Hofburg) was built continuously by the Habsburgs from the early 1200s until the early 1900s, and its total effect is of massive, burdensome display. But the individual parts open to the public (about half the building is used as government offices, and closed off) are rewarding to visit. These are:

The **Imperial Treasury** (Schatzkammer), where you can see the glittering crown of the Holy Roman Empire, made in the 10th century, and the crown, sceptre and orb of the Austrian Empire which replaced it in 1805, plus the dazzling treasure of the Order of the Golden Fleece and other riches.

The **Spanish Riding School** has one of the most beautiful Baroque, all-white interiors. On Wednesdays and Sundays, the elaborate leaps, jumps, gavottes, waltzes and other steps are performed by the gleaming white horses of the Lippizaner Stud. (Performances are booked well in advance.)

The **National Library**, whose Great Hall (Prunksaal) is a beautiful masterpiece of decoration, housing the book collection of Prince Eugene.

The **Emperors' Grave** (Kaisergruft), where three hundred years of Habsburgs are gathered together in their last resting place.

The **Chancellery** wing (entered from Michaelerplatz) houses the imperial apartments which are open to the public. They are filled with luxurious furnishings and tapestries, and contain the magnificent court collection of porcelain and silver.

The **Swiss Court** (Schweizer Hof, named after the Swiss Guard that used to be stationed here) was the starting point for the whole building. It was the site of the fortress where Ottokar tried to keep out the first of the Habsburgs, and in one corner of the Court stands the Burgkapelle (chapel), a little Gothic church, dating from 1449, where on Sundays one of the three teams of the **Vienna Boys' Choir** sings Mass.

The **Albertina** collection illustrates the graphic arts since the 14th century (drawings and engravings, especially Dürer).

The Environs

Some of the most enjoyable parts of Vienna lie outside the Ring. The two palaces of the **Belvedere** are just south of the city. The lower and upper Belvedere were the palaces of Prince Eugene, the lower his home, the upper his festive centre – Baroque masterpieces set in a Baroque garden. Today the lower Belvedere houses the collection of Austrian

Belvedere

Baroque art, the upper a collection of more modern Austrian art; the latter is the more interesting, reflecting and even predicting the decay and collapse of the Empire. In the Swiss garden (Alpengarten) is the museum devoted to 20th century art – Kokoschka, Klee and Klimt.

The cobbled hill that leads past the Belvedere begins at Schwarzenbergplatz, the most aristocratic quarter of Vienna outside the Inner City, where bankers and nobles were building their palaces, now infested with bureaucrats, right up to the eve of the First World War. To the right is the lovely **Musikverein** building, home of the Vienna Philharmonic, and also the **Karlskirche**, probably the finest Baroque church in Vienna (1716–39). To the

Karlskirche

left is the **Konzerthaus**, home of the Vienna Symphony Orchestra. The fountain at the head of Schwarzenbergplatz, just before the Belvedere, is a memorial (built by the Russians) to the Russian soldiers who 'liberated' Vienna in 1945, and were withdrawn ten years later following a treaty signed in the Belvedere. At the top of the hill is the South railway station.

The Palace of **Schönbrunn** is west of the city. Created as a retreat for the Empress Maria Theresia, it is imposing, magnificent, but not pompous. It is a Rococo palace, a country cottage on the Imperial scale built for the pleasure of an Imperial shepherdess. If you can visit only one of the sights of Vienna, this is the one. The most luxurious parts of the palace are the dazzling guest apartments, which include the Blue Room (Chinese tapestries), the Napoleon Room (Brussels tapestries), the Old Lacquer Room full of black-framed miniatures and the Room of a Million (Persian miniatures). These apartments contrast with the simple quarters where Franz-Josef lived, and the plain iron bed on which he died.

The gardens are even more attractive than the palace. They have always been open to the public, for Maria Theresia loved to be surrounded by 'her Viennese', and are laid out in a mixture of formal park with statues and fountains, and antique arbours. The Roman Ruins in the park were constructed there in 1778. The Gloriette, near the top of the park, is a classical colonnade which gives a lovely view over Vienna. The Wagenburg, at the side of the palace, houses a splendid collection of Imperial carriages, kept oiled, dusted and polished as though awaiting the Restoration. The Palmenhaus (palm house) is a stately home for tropical plants, and the zoo houses its animals in a Baroque splendour that is unique for a zoo.

To get to Schönbrunn from the city centre seems to start with a shopping expedition. There's a bus via the West railway station, which runs along the Mariahilferstrasse; this is the principal shopping street of Vienna, lined with the big department stores. Alternatively there is the train (Stadtbahn) which runs in a cutting alongside the Wien river. A good place to get the train is by the Naschmarkt, the biggest open-air market in Vienna. On Saturday mornings it is a flea market, overflowing along Wiener Linkzeile by Kettenbrucken station.

The region to the west of Vienna, **Wienerwald** (Vienna Woods) is on p. 51.

To the east of the Inner City is the **Prater**, a great pleasure ground dominated by the giant Ferris wheel, made famous by the film *The Third Man*. East too is the **Augarten park**, which contains the porcelain factory producing the china figurines you see in the city centre and the flea market. Beyond the main grey stream of the Danube, lined with wharves and warehouses, is the Old Danube (Alte Donau), a nearly blue backwater used for

Schönbrunn Palace

boating and sailing. Between the Donau and the Alte Donau is the 'U.N. city' – high, unappealing glass towers built to house branches of the United Nations, for in its dual role as a bridge and a neutral power, Vienna has become the third city of this international organization. Another international body whose headquarters are in Vienna is OPEC.

To the north are the garden suburbs of modern Vienna, which have almost swallowed up the wine-growing villages for which this district, generally known as **Döbling**, is still famous. The wine is still there, and the Viennese come out to Döbling to drink the new wine in an open-air *Heuriger* (p. 23); the most celebrated of these villages is **Grinzing**; **Nussdorf** is another, and **Heiligenstadt** is a place of pilgrimage for here, in 1802, in a house which is still a tavern, Beethoven, aware that his increasing deafness was incurable, wrote his *Heiligenstadt Testament*. (This is the best-known district for *Heurigen*, but there is also **Grossjedersdorf**, on the far side of the Danube opposite Klosterneuburg, which is the largest area with most music, and **Mauer** in the south.)

Beyond Grinzing are a few heights from which you can look out over the city while taking your ease. Nearest is the **Kahlenberg Hill** (483m/1570ft) which reveals the great tower blocks springing up round the edge of the town and the spire of St Stephans spiking up from the Inner City. Another is the **Leopoldsberg** (423m/1375ft) which also shows the city and gives a better view of the river.

Oberlaa park, in the south east of the city, is where you find the best flower displays, around the Kurzentrum; the Laaer wood next to it is more sparse. Near the park is the Zentralfriedhof (main cemetery) where most of the great musicians such as Beethoven, Schubert, Brahms and the Strausses are buried with suitable memorials. Only Mozart is missing – his remains lie in an unmarked grave in St Marxer Friedhof (St Mark's cemetery), by the junction with the south Autobahn on the way to Oberlaa.

EASTERN AUSTRIA

For most visitors the provinces of Niederösterreich and Burgenland are a playground extension of Vienna, for each of their points of interest can comfortably be visited in a day's outing from the capital. The Viennese themselves use these provinces at weekends for sport, scenery, or drinking, according to taste. But there is lots to see just for itself, without sight of the big city, with delights one doesn't usually associate with the picture of Austria as an alpine country.

Most un-Austrian looking is the Burgenland, around Lake Neusiedl which is used for sailing and bird-watching. This was part of Hungary until the collapse of the Empire when the German-speaking parts voted to join Austria, and the traditional costume as worn on Sundays, the food and the houses still seem to belong in central Europe. The land itself, east of the Lake, is the start of the great flat steppeland (*puszta*) that stretches for hundreds of kilometres almost to Budapest, dotted with ancient wells from which the water is drawn on a long pole. The hot climate of northern Burgenland (the southern parts are more raw) and black soil make this the garden of Vienna, rich in vegetables and sweetish wine, and famous as the home of storks.

Immediately west of Vienna is the Wienerwald, a mixture of tree-covered hills and terraced vines beloved by Viennese for soft, uncommercial charm and easy-going wine-drinking houses.

The prettiest part of the river Danube is that north of the Wienerwald, the stretch called the Wachau. A trip along the Danube by steamer through country

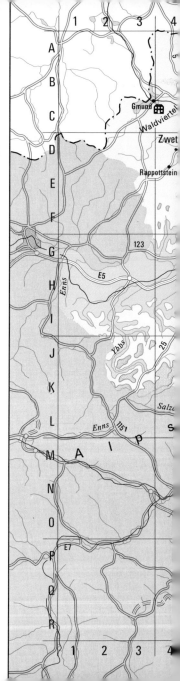

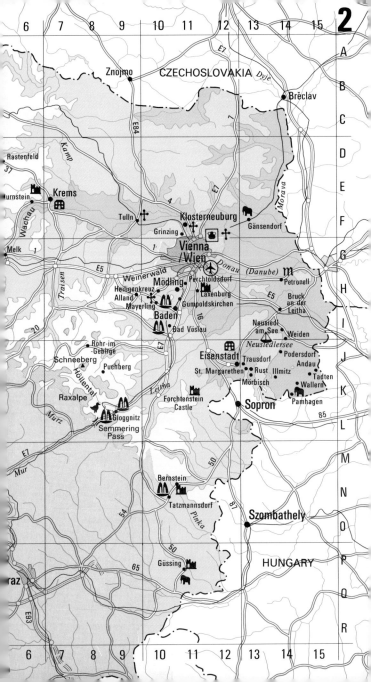

rather like the favourite parts of the Rhine is a popular summer outing, stopping off perhaps at one of the sleepy wine villages to visit a monastery or one of the castles whose ruins you see from the boat. North of the Wachau is the so-called Waldviertel (wood quarter), granite hills capped by dark pines, and fields of green pasture with oats between; this countryside is at its prettiest near the Czech border, where the valley of the river Thaya carves a way through the foothills of the Bohemian massif.

The Weinviertel (wine quarter) lies between Vienna and the Czech border, less rugged than the Waldviertel, producing some very good dry white wine, with many ruined castles and little towns where time has stood still. The Marchfeld, east of Vienna, is the granary of Austria; on the south side of the Danube opposite the Marchfeld are the excavated amphitheatres of the Roman capital of Carnuntum (at Petronell). Between the Marchfeld and the Weinviertel the vines are interrupted by the oil wells centred around Zistersdorf.

The only alpine parts of this region are in the southwest – the Rax and Schneeberg massifs before the developed resorts of Semmering, which are popular in winter for a day's skiing starting from Vienna, and the Ötscher massif (see Mariazell, p. 59) of limestone pre-alps, which is developing fast as a ski centre.

Pagan midsummer is celebrated in the Wachau, especially by Krems and Spitz, and at Ybbs-an-den-Donau; outdoor plays, operettas, concerts through much of the summer at Baden, Carnuntum, Klosterneuburg, Melk, Mörbisch.

September and October are the months for grape harvest and vinous festivities in all the wine-growing regions. Industrial fair at Wiener Neustadt in August. Winter sports begin at Semmering in December with a torch-lit ski-race, and as winter dies down there is Fasching (carnival) in most mountain places.

The most distinctive food is in Burgenland, which recalls its Hungarian connection with 'Pannonian cooking' – strongly seasoned soups, spicy fish.

Baden   H11

Niederösterreich (pop. 24,000) A typical and very pleasant spa town, centred on the casino and the hot springs which are used at the cure centre as warm baths or to warm a sulphurous mudbath. The park is rich in roses and used in summer for plays; theatre in winter. Centre of the town is a pedestrian precinct, occasional outdoor performances in the main square. Baden was very fashionable in the early 18th century; now it is mainly a refuge from Vienna, with which it is connected by tram.

Eisenstadt   J12

Burgenland (pop. 10,000) Eisenstadt, the capital of the province of Burgenland, is hardly more than a big village, with a single main street running down from the Esterhazy Palace to the little alley which leads to the house where Haydn lived while in the service of the Esterhazys. Some rooms and galleries in the palace are open to the public (except Sunday morning and Monday), especially the great hall where Haydn conducted his symphonies or played in his quartets most evenings for his aristocratic patron; Haydn's house is also open, and contains also mementoes of Liszt (born at Raiding, 46km/29mi south). Haydn's tomb is in the parish church.

Eisenstadt is important as a wine market, and the road east, towards Neusiedl, is called the Weinstrasse, lined with cellars and stalls selling wine and local produce like corn dollies, plants, and sub-tropical fruit.

Forchtenstein   K11

Burgenland The castle of Forchtenstein is set in romantically rugged country – chestnuts and cherries alternating with grim firs – and from the distance as from close up looks like something from a fairy tale. It is open all year round (about 3 schillings without guide, about 20 more schillings for guided tour). Built around 1300, with an armoury containing weapons from 16th to 19th centuries; light-hearted plays (*eg Zigeunerleben*) in summer evenings.

Gänsendorf   F13

Niederösterreich (pop. 4000) Chief town of the mini-Texas of Austria, but no Dallas – not one hotel. The oil wells are mainly to the north, safari park in the south. **Zistersdorf** is another oil centre, equally bare.

Klosterneuburg   F11

Niederösterreich The great abbey (open daily 0900–1200 and 1300–1800) is the chief attraction of Klosterneuburg – built with great luxury over a hundred years from 1730, modelled on the Escorial palace near Madrid, but incorporating many bits from the 12th to 14th centuries. The abbey is also the principal producer of wine in this rich wine-growing district of the Wienerwald, and sliding down its great cask on 15 November, name day of Leopold, founder of the monastery, is a simple amusement. The town is a favourite expedition centre for Viennese, and puts on an annual festival with symphony concerts and organ recitals.

Krems   E7
Niederösterreich (pop. 22,000) The Wachau is the picturesque stretch of the Danube between Krems and Melk, rich black earth terraced to support the southern fruits for which the region is famous and the vines that make Krems a small wine capital. The tasting of the season's new wines seems to start even earlier in Krems than around Vienna, and behind the plain doors there is laughter and the chink of glasses at any time of the year. It's an old town, four or five narrow streets terraced above the river, and instead of the street square typical of later towns there is a single main road (Steinerlandstrasse) lined with houses dating from the 1530s, massive and heavy, with little back alleys unchanged since then. Museum of wine in the old Dominican priory, and distillery fumes blending with the rich scent of the wine.

Dürnstein, 7km/4mi upriver from Krems, has a narrow medieval street running parallel to the river along a low ridge, and is famous for its castle, the **Kuenringburg**. The ruins can be reached in under half an hour along a stony path from the east end of the village. Here is the high cell where Richard the Lionheart was imprisoned by Duke Leopold V of Austria after a squabble during the Crusades. Richard was discovered by his faithful servant Blondel, and his ransom paid for many of the noble buildings that still decorate this part of Austria. Today the Castle Hotel (overlooking the river), the Minstrel Blondel Gasthof, the Richard Lionheart Hotel, and Blondel souvenirs, continue to enrich Dürnstein.

Across from Krems the monastery of **Gottweig** marks the lower end of the Wachau as **Melk** abbey marks the upper.

Melk   E7
Niederösterreich The great abbey of Melk, on a promontory overlooking a small arm of the Danube, is the finest example of Baroque architecture in Austria. The marble hall with painted ceiling is well worth a visit, and is linked by a terrace to the library of 80,000 books, even more worth visiting. But the greatest treasure is the lavishly decorated many-windowed golden church. The abbey was built in the early 18th century on a site which was the first capital of the Babenbergs when they became rulers of Austria in 976 – from here order and culture spread throughout the Austrian Danube.

Melk Abbey

Schallaburg castle with its richly decorated Renaissance façades is 4km/2½mi from Melk. Upstream from Melk, as far as the dam at Persenbeug, is the Nibelungen district, centred on **Pöchlarn** which claims to be the town of the Nibelungen. *The Ring of the Nibelung* is the name of Wagner's opera cycle which ends with the death of Siegfried, but in the original epic (a long poem written down at Melk in Middle High German) that was merely the beginning. After Siegfried was killed, his widow Kriemhild travelled down the Danube to meet Attila at Tulln and later persuaded Siegfried's killers to assemble here in the Nibelungengau before coming to Attila's court and extermination.

Neusiedl am See   J14
Burgenland (pop. 2500) This is the first village on Lake Neusiedl for the sailing and fishing enthusiasts who stream out of Vienna to the huge car park at the head of the lake. The village is a street of low, double-pitched houses – with a little less ornament it could be Irish – and on the outskirts the grapes and the chickens in the backyard begin at once. There is a lake museum.

Lake Neusiedl, 36km long and from 7 to 15km wide, is only two metres deep, a steppe lake. It is surrounded by a broad belt of reeds, which house a great variety of water fauna – herons, storks, bustards and over 200 other species. Warm enough in summer for swimming, but rather dirty; in winter there is ice-sailing. The south east side of the lake is a nature reserve, with a biological station supported by the World Wildlife Fund, and provided with observation towers.

The villages along the east side of the lake – **Weiden**, **Podersdorf**, **Illmitz** – are set back a kilometre or so from the shore and connected to it by roads through a wall of reeds. Quiet during the week, but at weekends many of the villagers wear local (Hungarian-style) costume while the townees come out in their Sunday suits to ride in horse-drawn charabancs down to the lake where they sing and eat, or ride on a lake steamer.

The villages away from the lake are true settlements of the plain – low simple houses compacted together as though afraid to venture into the wild steppe. Although not far from Vienna, they are isolated, quite unsophisticated, a relic of another age, but with simple fun at a street fair or beer tent.

In the background looms the Hungarian frontier. On the Austrian side the roads which used to cross the border are maintained as though expecting traffic, even with an idle customs post, but on the Hungarian side the road stops abruptly at a high earth dam, overgrown, land-mined, forbidden territory. Elsewhere on the frontier the Iron Curtain is just a low barbed-wire cattle fence, with a gap in it tied up with twine, and a well worn footpath through the gap, and a grim watch tower surveying the footpath.

See quiet **Pamhagen**, with its 'zoo' of steppe animals, all but the wolves left to roam free; or **Andau**, making the most of its only feature, a concrete-lined wet hole called the Puszta Lake.

Petronel   H14
Niederösterreich (pop. 1400) Just south of the Danube, centre of the excavations of the Roman city of Carnuntum which was the capital of Pannonia – you can still taste the 'Pannonische Küche' in Burgenland. Extensive ruins, good museum, plays are presented in summer in one of the several amphitheatres.

Rust   J13
Burgenland (pop. 1700) A former Imperial free city – granted the privilege because of the quality of its wine – Rust is still graced with solid, arcaded houses of the vintners, some of them richly decorated in a Baroque style, and is also home to hundreds of storks who nest on all the pinnacles, and to basket weavers whose produce from the local reeds is on sale. A road leads through the reeds to the lake for water sports and fishing from punt-shaped rowboats.

South of Rust, **Mörbisch** is another small town famous for its wines, tight on the Hungarian frontier. Swans waddle across the road down to the lake, where there are the usual water sports and a floating stage for occasional theatre.

At **St Margarethen**, on the road from Rust to Eisenstadt, there is a working quarry which has been used since Roman times, and provided much of the stone for Imperial Vienna. Here there is a permanent exhibition of the work of sculptors who use the Sandkalkstein from the quarry, and from May to October there is a passion play on Sundays. **Trausdorf**, beyond St Margarethen, is one of the more colourful of the wine villages of this region, with many sellers lining the roadside.

Semmering   L9
Niederösterreich (pop. 1000) The Semmering pass, on the densely trafficked road between Vienna and Bruck marks the turn-off to the international resort of Semmering, which is an array of little, big, and giant hotels and guesthouses strung out along the tree-lined road that winds

down from the pass to the railway station. It is busy in summer as a high altitude 'Kurort' – fresh air in place of mineral springs – and in winter as a ski centre, giving access to the limestone massifs of the Raxalpe and the Schneeberg which are separated by the Höllental (Hell Valley). This valley is not so dramatic as its name would suggest – a gentle winding slope beside the little river Schwarza. In this region, **Reichenau** is a spa-like summer resort, with cable car direct into the Rax, and vestigial mining centre; **Puchberg** is a sleepy village with a rack railway direct into the Schneeberg; **Rohr-im-Gebirge** is scattered in a level valley surrounded by high mountains, where there are still plenty of deer; **Gloggnitz** is a former mining town which has been smartened up into a mountain spa, with parks and wood-walks and mudbaths.

Bad Tatzmannsdorf   N10

Burgenland A modest spa, with parks and cure-hall based on mineral springs and 'curative bog', Bad T is a handy centre for the southern Burgenland – an intensely cultivated stretch of small fields, overlooked by mountains rich in castles. The richness of the fields and the peaks on which the castles stand show that this was once a land of volcanoes. In the immediate vicinity of Bad T are the open-air museum (open all year round) and the medieval castle of **Schlaining** (open Easter – October, closed Mondays); further afield are **Bernstein**, with its castle and its workshops selling vases and jewellery in the local jade, and **Güssing** with its 12th-century castle (open daily April – October) and wild animal park.

Tulln   F9

Niederösterreich (pop. 8000) The level plain around Tulln is a noted flower-growing region, and there is a regular fair devoted to horticulture. The woodland is rather scrubby, but the school of experimental forestry is interesting for serious silviculturalists. The town itself is mainly 1920s, with the Kärner (funerary chapel), dating from 1250, of interest to specialists.

Waldviertel   C5

Niederösterreich The Waldviertel (wood quarter) is the mountain-free country east of Linz, between the Danube and the Czech border – no more wooded than many other parts of Austria, a land of chequered plains and forested hills, with many small lakes, gentle rivers, untouched villages, and ruined castles. The chief towns are **Zwettl**, which is a market town with a beautiful Cistercian abbey, splendid library, and ready access to the castles of **Ottenstein** (in the lake), **Rastenberg**, **Rappottenstein** (nice little market town by the river Kamp) and **Rosenau** (Baroque palace). **Gmünd**, on the Czech border, is a glass-making town with noted sgraffito-decorated houses; castle **Heidenreichstein** is still moated; viewing tower in the geological open-air museum; nature reserve at **Fibenstein**. **Retz** is now an important wine centre (tour of the cellars) but has a quaint old town, regularly laid out, with the town wall virtually intact.

Wienerwald   H9

The region immediately to the west of Vienna is called the Wienerwald, a mixture of limestone hills, gentle wood and some dense forest, and intense cultivation of the vine; despite the romantic associations of the Vienna Woods (Tales of), it is no more wooded than many another part of Austria, but forms a lung for Viennese to breathe. Some visitors to Vienna stay in the Wienerwald for economy. The vintners' villages are full of Heurigen (p. 23) where the vintners sell their own wine to be drunk in their orchards or courtyards, sometimes with good food, always with good company. Most true-to-form of these villages is probably **Gumpoldskirchen**; **Perchtoldsdorf** is another, larger, well-loved drinking spot; **Mödling** preserves the name of a wine town but has become a substantial industrialized 'Europastadt'; **Bad Vöslau** is off the regular imbiber's route but produces a very fine light red wine.

For pure woodland walks the little village of **Wienerwald** itself is a good centre; other places of interest are **Mayerling**, where there is a convent marking the site at which the Archduke Rudolf, heir to the Austrian throne, shot his young mistress and then himself in the 'Tragedy of Mayerling' in 1889; **Heiligenkreuz** with the beautiful Cistercian abbey (to which the chief of police rushed the girl's body for hiding while he tried to cook up a story to tell the Emperor that Rudolf was dead); **Alland** with a stalactite cave; the castles of **Laxenburg** (Baroque, with a very Baroque park) and **Liechtenstein** (Romanesque), both near Mödling; at **HinterBruhl** is an old mill which allegedly inspired Schubert to write one of his greatest song-cycles ('Die Schöne Mullerin') and also an underground lake, largest in Europe, where visitors can sail.

The prettiest stretch of road is the Helenental, running down from Mayerling to Baden along a little river through beech woods, but really you don't need to select any one place in the Wienerwald for beauty; all of the south is pleasant.

SOUTHERN AUSTRIA

The Tauern range falls away eastwards from the Grossglockner and Gross Venediger, Austria's two highest mountains, cutting off the well-known Alps of the Tirol and Salzburg from the southern provinces Osttirol, Kärnten and Steiermark. This southern side of the range is less developed for tourists – in its more mountainous parts there are quiet farms and little hamlets still leading their rural life, with no brochure or claim to be a resort; just a modest Gasthaus visited by the lucky few who already know it.

The southern side of the range is noticeably warmer, too and this is most apparent in the lakes of Kärnten (Carinthia) which lie along the broad valley of the river Drau. These deep clefts surrounded by mountains are lined with little summer resorts, popular for swimming and boating. The rest of the Kärnten is alpine – the high Alps of the Tauern, or the lower limestone of the rugged Karawanken.

The much larger province of Steiermark (Styria) extends all the way from the classic alpine scenery of the little part of the Salzkammergut that is in Steiermark, in the northwest, down to the open plain around Radkersburg, the most southeasterly town in Austria; the Alps get lower and lower as you travel southeast and fade away after the provincial capital, Graz.

Steiermark likes to be known as the green mark of Austria, for it is the most heavily wooded province, but could as well be white Styria, for there are nearly 600 ski lifts, mostly in unsophisticated centres, as well as the Dachstein and Mariazell areas. Or it could be red Styria for the iron ore that provides the industrial wealth of the area around Leoben. Steiermark is a great province for game, with herds of deer in the wildlife parks of Mautern (Liesingtal), Alpl and Herberstein.

Styrians cling more fervently to their old ways and loyalty to the province rather than the country – a characteristic which is reflected in their low key approach to tourism. The Archduke Johann is still lovingly remembered and you will encounter his name everywhere. As a son of the Emperor Leopold II he was made Governor of the province in 1810 and devoted his life to building up its railways, industry and learning – apart from the escapade for which he is best loved, when he ran off with the daughter of the village postmaster of Bad Aussee. They use two adjectives for the province; Steirisch suggests modern Styria, which condescends to be part of Austria, while Steiermärkisch has an old-fashioned ring, recalling the grand days when it was an outpost of Empire, a line of fortresses against the Turks or even a 'march' of Charlemagne holding off the Huns.

Remembrance of the past is even stronger in Osttirol (East Tirol) but here it is the more recent past. Under the Empire Tirol was a vast province, stretching almost from Verona in Italy, but when Italy claimed the Italian-speaking parts in 1918 the frontier was pushed up to the Brenner pass and the German speaking majority of South Tirol was included in Italy. This created a sore which is still running and cut off East Tirol from its big brother, Tirol proper, or North Tirol.

Under Hitler Osttirol became part of Kärnten for administration, and this arrangement continued under British occupation, but when Austria recovered its independence in 1955, the ties of brotherhood and history were stronger than the logic of geography and Osttirol is now governed again from Innsbruck.

The chief reason for the Felbertauern toll tunnel was to provide a short, all-year road connection between North Tirol and Osttirol, and it has had the effect of opening up the valleys of the area to tourists. The mountains are strikingly rugged, barely softened by cultivation and in the vicinity of the Gross Venediger and the Dolomites south of Lienz developing as a ski centre.

Admont E13
Steiermark (pop. 3000) A quiet little village with a Benedictine abbey whose chief attraction is the library which escaped the fire of 1865 and can be visited all year round until at least 1600. Splendid state-

room (late 18th-century) with painted ceilings and woodcarvings and about 150,000 books. A popular outing, flower sellers by the gates. East of Admont the river Enns tumbles through a gorge called the **Gesäuse** which is a favourite for rough-water canoeists and for fishermen. The foaming blue water, surrounded by bare crags is a sight for landscape lovers too. An especially attractive detour is up a side valley to **Johnsbach.**

Bad Aussee E10

Steiermark (pop. 5000) A high-altitude spa, with parks and gardens, shut in by bare mountains which also keep off the wind. Centre of the old town is the **Chlumetzkyplatz**, cobbled and rather decrepit, with a couple of interesting Gothic buildings and the classical-style offices of the spa administration. The spa is mainly for digestive treatment and mud-baths. Nearby **Lake Grundl** is a miniature Lake Luzern, with vistas of criss-crossing mountains, green slopes on the far side backed by cliffs, overlooked by the bleak Totes Gebirge.

Bruck an der Mur G18

Steiermark (pop. 16,000) Visited because it's an important road junction, the quiet centre of Bruck, of 16th-century houses round the Hauptplatz, especially the house of Pankraz Kornmess, is an escape from the surrounding iron industry. Easy walks into gentle countryside, *eg* the path to the 'Cold Spring', or more strenuous walk along the wooden platforms through the gorge of the **Bärenschutzklamm** at Mixnitz, 12km/ 17½mi long.

Donawitz G17

Steiermark Music lovers make their pilgrimage to the homes of their heroes in Salzburg, Eisenstadt, Linz or Heiligenstadt. If your interests are more in industry, where better than Donawitz, birthplace of the oxygen-blown converter that revolutionized steelmaking in the 1950s. There's no memorial to it, and not much else in this wire-drawing suburb of Leoben (pop. 35,000). Leoben, home of the famous Goss beer, a metallurgical university, a well preserved old theatre and modern congress facilities, offers the tourist little beyond a handsome main square, the ancient gate by the river bridge, and the atmosphere of serious enjoyment. But the road north of Leoben is well worth taking. Its chief interest is the **Erzberg** – the Ore Mountain – a 1500m/4920ft high pyramid of pure iron ore which is continuously mined at the surface from a remarkable series of over 30 terraces, like a ziggurat. The workings and tunnels can be visited from the village of **Eisenerz** (tours start at 1000 and 1430). On the way to the Erzberg, mining settlements like **Vordernberg** have their 19th-century workers' housing blocks, very similar to their parallels in the industrial north of Victorian England, yet with an Austrian amusement, as though the Empire did not really believe in industry. Beyond Eisenerz the country is softer – it can be seen as a whole by chairlift from **Präbichl** to the Polster – with a little detour to Lake Leopoldstein whose deep green waters are darkened by the high cliffs of the Kalte Mauer – the Cold Wall. Trout is sold by the roadside – whether it comes from the ore-red rushing stream you are not told. Legend has it that a merman was trapped in the Lake, and a ransom offered. Gold for one year, silver for ten, or iron for ever. His captor chose the iron, and thus the mountain was converted to the ore, which will last 150 years at current rates of extraction.

Graz J19

Steiermark (pop. 248,000) You need to poke around in Graz to discover many of its treasures. The outskirts of the city are heavily industrialized, and associated with the Graz 'Southeast Fair', an engineering and production display held in May and around the end of September. Surrounding the city is a ring of high-rise flats, and within that are suburbs of 6- and 8-storey blocks – blocks being the operative word. But within these is the old town, which is not a small left-over like most 'old towns' but the full size of a medieval city. A map and guide, obtainable free from the tourist office, lists and illustrates the sights, but there are many others to be found just by wandering around.

The first problem, if you arrive by car, is to park – it's hopeless to try to drive round the inner city. If you just follow the main stream of traffic from one of the bridges over the river Mur, there's an underground car park on Andreas Hofer Platz, other car parks on the edge along the Opernring and Burgring, or if you nose your way in from the Burgring you can, if you're early, find space on Freiheitsplatz.

On foot the principal street is the **Herrengasse**, which leads south from the **Hauptplatz** and is the most elegant shopping street, strong with the smell of coffee. In the Herrengasse is the **Landhaus** (built about 1560), former seat of the Styrian parliament, with superb arcaded courtyard and inner chambers, and next door is the **Zeughaus** or Arsenal, which you must visit. In the 17th century when

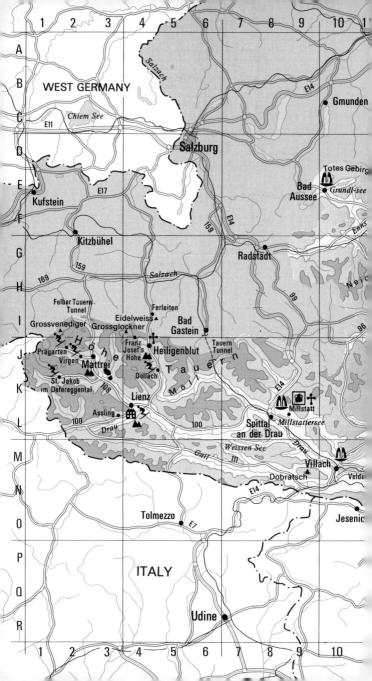

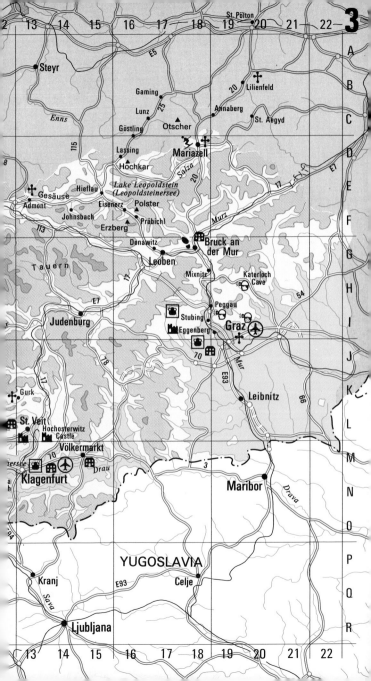

the Turks still threatened the city the armour and swords and pikes and muskets for an army of 15,000 men were kept here, and whenever there was trouble the citizenry was armed and sent out to defend the city. Then the arms were collected, cleaned and oiled and stored in the arsenal until the next invasion. Since the last time they were used, in the early 1700s, the weapons have stayed here, gleaming and rust free, stacked on four highly polished floors – the largest collection of such weapons in the world. Open 0900–1200 and 1400–1700.

North from the Hauptplatz runs the narrow **Sackstrasse**, which leads to the funicular railway that takes visitors to the Schlossberg. Along Sackstrasse there are many restaurants, tucked away in the courtyards which you approach through arcades on the street that conceal the riches behind. There are some grand houses too, unmarked to say what they are, like No. 17, directly opposite the steps down from the Schlossberg, with some exceptional murals and ceiling paintings, and others with statuary, aristocratic staircases, classical furniture – all unidentified, but nobody minds if you just push a door open.

The **Schlossberg** is the hill overlooking the city where the fortress was until the Napoleonic wars. The French destroyed the castle but were paid by the townsfolk to spare the clocktower and belltower, out of sentiment. Today the castle hill is a series of gardens and chestnut terraces.

The back streets between Sackstrasse and the river and off the **Sporgasse** are a medieval maze of shops, some now modernised into boutiques, others still open-countered. Next to the Cathedral the former **University** should be seen for its red Renaissance courtyard and the busy life of the Jesuit college now resident there. Even the 19th-century government buildings, near the Post Office, are a revelation of grandeur, red marble, and wasted space.

The **Joanneum** is a museum of Styrian history, industry, and the arts founded by Styria's favourite son, the Archduke Johann, and also houses the **Old Gallery**, a collection of paintings, and fine stained glass; the **New Gallery** of more modern paintings is in the Sackgasse. Theatre, opera house, three universities, concerts in the lovely **Minorites Hall** (next to the **Mariahilf church**) in Mariahilfestrasse across the river), and in the Stephaniesaal.

Just outside Graz are a number of worthwhile visits. **Schloss Eggenberg**, 3km/1mi west, is an Italian-looking palace of the early 17th century with magnificent

Clocktower, Graz

decorated state rooms revealing Baroque taste of the late 18th century. The Austrian Open Air Museum, at **Stubing**, 16km/10mi north of Graz near the start of the motorway (open April – October, closed Mondays, last adm. 1600) houses rural buildings of all the main styles found in Austria, together with their traditional household equipment, in settings as near as possible to the original. The **Katerloch** (Kater Hole), about 35km/22mi north of Graz, which the fit and well-shod climb down by ladders, has a giant hall of stalagmites and stalactites and an underground lake. The Lur grotto, at **Peggau** beyond the Open Air Museum (closed Mondays), is similar but less impressive, with an underground river. Animal park at **Herberstein**.

At **Piber**, 30km/18½mi west of Graz, there are afternoon tours to the stud farm where the Lippizaner stallions are bred for the Spanish Riding School in Vienna.

Gurk K13
Kärnten (pop. 1300) Gurk Cathedral, completed in about 1200, is the finest example of Romanesque architecture in Austria, plain arches, vaulted chapel with 700-year-old murals, and a crypt supported by columns of Doric simplicity. The decoration is mainly Baroque – gilt ornamentation and exuberance, shown up by the majesty of the stone.

The adjacent priory buildings, plain arches from the 15th and 17th centuries, show how architecture could be refined without becoming ostentatious.

Heiligenblut J4
Kärnten (pop. 1500) Heiligenblut is noted for its church, perched on the edge of a grassy mound, with its spire pointing up the distant view of the Grossglockner (3797m/12,457ft); as the winter headquarters of the famous Heiligenblut

Heiligenblut

mountaineering school; and in summer as the starting point for expeditions into the mountains and glaciers of the Hohe Tauern. The village itself is now just shops and hotels, always packed with tourists, squeezed between the mountain and the valley car parking at the end away from the main road. **Gössnitz** waterfall is beyond the car park.

The road across the Hohe Tauern north of Heiligenblut is a toll road, open May to November. Principal attraction is up the spur road (may be closed longer than the main road) to the **Franz-Josefs-Höhe**, furthest point in the mountains reached by the old emperor when he first explored the region. From here mountaineers walk over the snows to the Archduke Johann hut (3454m/11,321ft) for the ascent of the Grossglockner while sightseers take the funicular to the **Pasterze Glacier**, 10km/6mi long. The more energetic follow the path for the Oberwald hut (2972m/9750ft) for the view across the glacier from Wasserfallwinkel. Further north on the main road, a favourite panorama is from the observation tower at **Edelweiss-spitze**; the ravine at **Piffkar**, above which there are splendid mountain views, and the falls in the **Walcher** rivulet (north of Ferleiten) are other attractions.

Whereas the Felbertauern Alpine Road

Grossglockner and Pasterze Glacier

is rather bleak, with uncultivated rocky hillsides nearby and scree-covered mountains in the distance, the Grossglockner High Alp Road which takes in Heiligenblut is very agreeable even in the mist. Green hillsides with little villages dotted here and there, and vistas of interlocking snow-covered peaks, together with some awesome bends and corniche roads, and a barren sinister stretch between the two tunnels. Down in the valley, **Döllach** is a little village of sharp-bending streets with a smart new museum in a ramshackle little castle, and some winter skiing.

Hochosterwitz   L13
Kärnten Probably the most fairytale-like of all Austrian castles, standing on a conical hill high above the plain. From the car park it is about ½ hour walk up the ramp, past gateways and drawbridges, to the castle itself – open 0900–1800 May to October; collection of weapons and armour.

Hochosterwitz

Klagenfurt   M13
Kärnten (pop. 85,000) You have to find your way, peering through private-looking doorways into arched courtyards, to discover the past riches of Klagenfurt, largest town and capital of the province of Carinthia. The **Alter Platz** (closed to traffic) is a street-square, still used as a market, widening towards its grander end where the **Landhaus** with arcaded courtyard and superb staircase is the principal monument. Off the Alter Platz are many narrow streets and alleys, especially the **Herrengasse**, lined with houses of the nobility, now broken up into professional offices but still showing the formal, grandiose interiors of the days when dukes and counts – and Napoleon – lodged there. The **Neuer Platz** is the modern centre of Klagenfurt, a rectangle surrounded by traffic; the modern stone dragon (Lindwurm) in the middle, which is the symbol of Klagenfurt, is the fruit of medieval imagination following the discovery of a rhinoceros skull which is now on display in the regional museum. Very lively little market in the **Benediktiner Platz** (try the carrot juice at 0600); concert hall; provincial-looking theatre; industrial fair.

The camping site and eastern edge of Lake Wörth are 3km/1.8mi outside Klagenfurt. Here are the outdoor amusements and the vast, gloomy, bathing establishment for swimming in the warm waters of the lake. Boating round the south shore of the lake. Nearby **Minimundus** is a miniaturized world for child and adult visitors

Minimundus, near Klagenfurt

Lienz L4
Osttirol (pop. 12,000) A quietly busy little town, with a southern relaxation, not destined for tourism but still filling up with knowing summer visitors. There are two main squares, one a market place by the river connected by arcades to the other, the **Hauptplatz**, which is a pedestrian precinct with cafés along one side where you can sit in the sun and admire the **Liebburg**, a twin-towered 16th-century palace which now houses government offices. At the edge of the town is **Bruck Castle** (open 1000–1700 July – August, otherwise closed Mondays and from Palm Sunday to All Saints' Day), a mixture of dilapidation and restoration, with an interesting and bare medieval Knights' Hall, Gothic chapel with 15th-century frescoes, and museum – and softly wooded grounds around a small lake. Interesting collections of minerals from the nearby Dolomites at the castle, and on sale from stalls by the roadside, for these sharply gouged mountains attract many amateur geologists. 14th-century painted *Siechenhaus* in the **Kärntnerstrasse**. Popular in summer as an excursion centre, just waking up to the possibilities of winter skiing, with direct access to the Austrian Dolomites.

Wild game park at **Assling**, on the road southwest up the Drau valley; cable car to the sunny plateau of **Zettersfeld**; almost daily orchestral or brass band concerts in July and August.

Maria Wörth N12
Kärnten (pop. 1000) A small resort on a promontory, noted chiefly for its beautiful pilgrims' church, 12th-century, approached up roofed wooden steps. From **Reifnitz**, a small version of Maria Wörth just outside it, the road goes up to the 900m/2950ft high hill called Pyramiden Kogel, where there is a high viewing tower by the restaurant.

Countryside in Steiermark

Mariazell D18
Steiermark (pop. 2300) Mariazell is famous as a pilgrimage centre, for the basilica housing its gold-crowned Madonna under an ornately silver dome was a cult centre under the Habsburgs, and there are still torchlight processions on summer Saturdays. The little town, compact between the fir woods and the hayfields, is popular with summer visitors as a fitness centre for the active, and in winter is a skiing centre for the Viennese. 5km/3mi outing to the **Marienwasserfall** (waterfall of Mary).

Mariazell is also an expedition centre for visitors to the **Ötscher massif**, in Niederösterreich. This is gentle pre-alps leading up to the rugged peak of the Ötscher, with good skiing. Chief village is **Annaberg** (waterfall at Lassing, rocky streams); others are **Gaming** (Carthusian monastery, road to nature park), **Göstling** (for the Hochkar plateau) and **Lunz** (pretty lake) – all fairly unsophisticated, folklorey, comfortable, developing fast as ski-region. The St Aegyd valley is more industrialized. At **Lilienfeld**, further down the valley in the direction of St Pölten, is a Cistercian abbey of the Babenbergs with a huge, clean-lined church of the high Middle Ages.

Matrei in Osttirol J3
Osttirol A mountain village just off the main road from Lienz to the Felbertauern tunnel, made interesting to look at by the fact that it is on a slope, not in a valley. Popular summer mountaineering and

walking centre, being within sight of the Gross Venediger range which reaches as high as 3674m/12,053ft. Immediate access to the mountains from the Matreier Tauernhaus, 16km/10mi away at the entrance to the tunnel. Two villages further up the valley from Matrei which are more typical alpine villages and used as winter sports bases are **Virgen** with some pretensions as a spa because of the clear warm atmosphere and **Prägraten** near little Lake Berger and the Umbal waterfalls.

Don't confuse this with **Matrei am Brenner**, about 23km/14mi south of Innsbruck which is a recently re-built settlement of houses in the old Tirolese urban style, and more of a rambling centre, with some very low-cost accommodation.

Millstatt L9

Kärnten (pop. 1300) A resort on the Millstatt lake, the Millstattersee which is very deep but very swimmable – it reaches 26°C in summer. By the lake there is a salt-spa, and next to it hotels and restaurants crowded on top of each other, but the rest of the town is cool with an airy, cobbled modern market place above the ancient abbey. The sunny, yellow-arched outer courtyard of the abbey (now administrative offices) contrasts with the cool white of the inner cloisters where there is a museum recalling the Order of St George, founded when Millstatt was a frontier town against the Turks.

Spittal (an der Drau) L8

Kärnten (pop. 1300) A modern, clean, pleasant town, quiet although it is a main crossroads. Very green public park just off the main road, formerly the park of the dreamy Renaissance Porcia Palace (0900–1800 May – September) whose arcaded courtyard is used for concerts and comedies.

If you drive to Spittal from Lienz or Badgastein, a detour through the green, level Möll valley is worthwhile to visit the Ragga Gorge – you walk up the cleft above the water on a wooden ramp, and then back down along a forest path – less water and more noise than the Aare Gorge (p. 101) or the Tamina Gorge (p. 111). Cable car from Spittal to the Goledk ski-fields.

St Jakob im Defereggental K2

Osttirol St Jakob, like the other villages of the Defereggen valley which lies between Matrei and Lienz, has its ski lifts and toboggan runs to take advantage of the winter snows, but remains essentially an unspoilt – and pretty – mountain village that enjoys itself and enjoys visitors.

St Veit an der Glan L13

Kärnten (pop. 13,000) A medieval city, former capital of Carinthia, modernized in the 18th century so that it has two 'street-squares', and is now surrounded by pleasant suburbs living off the cloth industry. The lower square is a cobbled street lined with restrained, plain-fronted burghers' houses, now shops, and leads through an interesting portal (see how the owners of two separate houses have co-operated to make their lines match and compose the portal) into the main square. The broader main square, also cobbled and with a fountain whose basin is Roman, is a lesson in architecture. Its chief building is the **Town Hall**, in the middle of the long side, a Gothic building with a black-and-white tiled courtyard and vaulted great hall which was given a new façade in the 1750s in the Baroque style – compare its lavish pediments with the simple ornamentation of the **military headquarters**, only thirty years later, in the west side, and with the later hodge-podge of styles in the façade of the savings bank on the east side.

St Veit is the centre for excursions to a large number of castles; **Hochkraig**, and **Niederkraig** are about two miles (3km) walk from the town; **Frauenstein** is very well preserved, about 5km/3mi by car and 1km/1½mi on foot; **Hochosterwitz** see p. 58.

Villach M10

Kärnten (pop. 35,000) Mainly an industrial town (timber) and communications centre, Villach turns its back on the dark river Drau and is centred on an old square facing the parish church and modern pedestrian precinct. In the Schiller Park a huge panoramic relief of the province gives a clear picture of the mountains you may wish to visit. South of the town is the spa centre, Warmbad Villach, with hotels and cure-establishments laid out in restful parkland to take advantage of hot springs which since Roman times have gushed out radon-rich waters for relief of rheumatic and circulatory disorders. Good outing to the wooded Villach Alps and cable car to **Dobratsch** for 2 hour walk to splendid panorama.

Völkermarkt M14

Kärnten The principal street, leading at right angles to the main road through an 18th-century arch, is an attempt at a street-square with residual elegance in some of the buildings, and in the back streets are many more once-fine houses and small workshops in faded yellow. A market struggles on in a small square off the main street. Rock bottom prices.

NORTHERN AUSTRIA

Beautiful lakes in the clear air of the mountains, land of the White Horse Inn and the Sound of Music, ski-slopes and spa-parks, woodland cafés and elegant palaces – that is the popular picture of the Salzkammergut which is the heart of the provinces of Salzburg and Oberösterreich which make up our region of Northern Austria. A true picture, too, even though the words could be applied equally well to most of the Alps of Austria and Switzerland. Salzkammergut means 'domain of the salt office', for the district grew rich and comfortable on the labour of the salt mines, and the musical city of Salzburg, probably the greatest single tourist draw of anywhere in this book, was beautified on the proceeds.

Apart from its Salzkammergut, the province of Oberösterreich (Upper Austria) is mainly undramatic agricultural land. To the north of Salzburg city, between the river Inn and Danube, lies the relatively flat Innviertel (Inn quarter) which was acquired by the Habsburg Empire from Bavaria only in 1777, as compensation for territories lost to Frederick the Great. The stables of the large farmhouses are converted now to garages, but otherwise life goes on much as in simpler times, with perhaps more spontaneous music and dancing than in the salty Land of Music itself. East of the Innviertel is the more hilly country called the Mühlviertel (quarter of the river Mühl), almost wholly free of industry, with villages sleeping on hilltops and the ruins of castles and chateaux; high enough, in places, to have rudimentary ski facilities.

The Danube valley marks the southern edge of the Mühlviertel, its open space dotted here and there with large, manorial farmhouses set square round a central courtyard which recall when this countryside, from which Austria grew up, was broken up into huge estates; extensive infilling of modern buildings. Linz on the Danube is the modern industrial capital of Oberösterreich, Enns, just off the Danube, is Austria's oldest city.

The city of Salzburg and its part of the Salzkammergut are in the district called the Flachgau – the 'flat district', but only in comparison with the others. Most original of the rest is the Lungau, around Tamsweg, dry and sunny but cold, where folklore survives most strongly. The Pongau, centred on St Johann, and the Pinzgau in the west, centred on Zell am See and Lofer, are high-alpine areas, which have been, as the brochure so disarmingly puts it, 'cultivated by the inhabitants into an ideal holiday region' – that is, an abundance of ski facilities, sport, glorious mountains, accommodation of all grades, entertainment, but still not enough for the million visitors a year who come here.

Badgastein P5

Salzburg (pop. 6000) The pre-eminent spa of Austria, Badgastein has turned to being a winter sport centre as well, with readily accessible slopes in the 2100m/7000ft range, and an annexe 3km/2mi away, **Sportgastein**, which provides skiing in the 2500m/8000ft range. It is as a spa and health resort that Badgastein is most interesting. The town lies in a warm valley with high mountain air but sheltered from the wind, and still retains the flavour of 'taking the waters' in the high days of the Empire, though the shops are elegantly modern. A little stream rushes through the town, with a waterfall right in the middle, and either side are the greenery and conifers of a mountain spa. The hot springs are rich in radon which, when one drinks or bathes in the water at one of the hundred or so hotels, is supposed to stimulate the endocrine system. The favourite walk, quite level, is the Kaiser Wilhelm Promenade, with lovely views; this connects to the path along the mountain edge which leads to the less spa-like but less expensive resort of **Bad Hofgastein**. Bad Hofgastein in fact receives more visitors than Badgastein and together they welcome more even than Salzburg.

Freistadt D13

Oberösterreich (pop. 6000) Chief town of the Mühlviertel, a district of steep green hills with green pastures alternating with dark fir forests and granite outcrops. The main square of Freistadt, lined with softly painted old houses, the parish church and the Baroque town hall, is very restful, with quaint alleys leading to the old castle and remains of the fortifications. **St Oswald**, 8km/5mi east of Freistadt, on a gentle slope up to the parish church, is a very pretty village.

Gmunden I9
Oberösterreich (pop. 13,000) A lakeside resort, with a long, traffic-free promenade under the trees, swans by the concrete path, and – if the enthusiasts' fight to preserve it is successful – an old paddle steamer, the *Gisella*. The castle on the little island is floodlit at night.

Traunkirchen

Grein F15
Oberösterreich (pop. 2700) A minor resort on the Danube, at the start of the picturesque stretch of the river called the Strudengau. Small, unspoilt town square, and arcaded Renaissance courtyard of the Greinburg. Rocks and waterfalls at nearby ravine called the **Stillensteinklamm**. By road along the north bank of the Danube there are pretty little villages every few kilometres, some shut in by trees, some with open views across the river.

Hallein K5
Salzburg (pop. 15,000) A strong sense of history and wealth of things to see make up for beauty in and around Hallein. The nearby salt mines have been worked since neolithic times; they are still in operation, and open to visitors (0800 to 1700 May to September, by cable car from Dürrnberg) with a long walk through the shafts and galleries culminating with a slide down the toboggan run. The **Dürrnberg** was a centre of Celtic culture, and finds from that era are displayed in the **Keltenmuseum** in Hallein, housed in the 18th-century administrative block of the salt

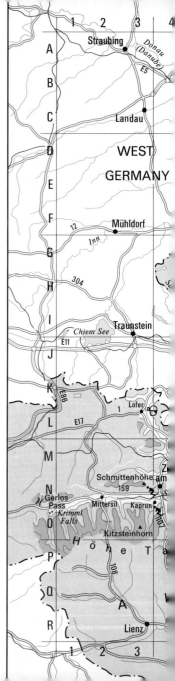

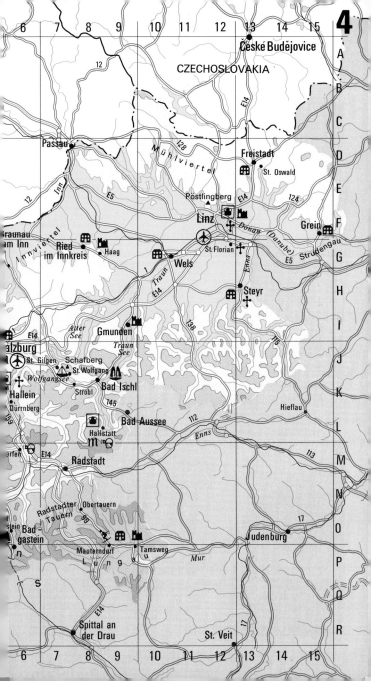

industry. In the museum, too, is a collection of mementoes of Franz-Xavier Gruber, who composed the carol *Stille Nacht, Heilige Nacht* (Silent Night, Holy Night) on Christmas Eve, 1818. He was organist of the parish church of Hallein where his grave is. The town reached its greatest prosperity in the 17th century, and the tall houses of this period with their steeply battered walls line the maze of narrow streets and little cobbled squares of the old town. The outskirts of the modern town are prosperous from the chemical and paper industries and have attracted a considerable Turkish population.

South of Hallein, turning off at **Werfen**, is the 'World of the Ice Giants', a system of caves where natural ventilation ensures that water freezes to fantastic architectural forms, with huge galleries and caverns and delicate tracery, domes, frozen waterfalls, statue-like columns – the largest such complex in the world (42km/26mi in total); magnificent mountain views on the approach road. Open 0930 to 1630 May to September; take warm clothing and stout footwear.

Hallstatt L8
Oberösterreich (pop. 1500) The oldest still-inhabited village in Europe, full of both interest and charm. Hallstatt is compressed onto a narrow terrace between the deep, black waters of its lake and the steep mountain behind, where salt is mined today as it was by pre-Celtic tribes of 1000 BC. The approach road today tunnels through the mountain, and opens into car parks from which you walk down past waterfalls to the pedestrianized village. The **Market Square**, a triangle of untouched 16th-century buildings, is a delight, as is the 18th-century lakeside terrace. The earliest Iron Age culture, from around 800 to 400 BC, is called the Hallstatt culture after the discoveries made here (just as the later Iron Age is called La Tène after the discoveries made on Lake Biel – p. 123) and many of the finds can be seen in the **museum**. Salt working at Hallstatt died out at the end of the Middle Ages but was revived under the Habsburgs and continues today in the Salzberg which can be reached by funicular; at the top the salt workings can be visited (including rides on a toboggan slide and rail-trolley) as can the Hallstatt-Age grave excavations.

From Hallstatt a winding road leads through pine forests with glimpses of bare mountain to the **Koppenbrüllerhöhle**, a massive system of underground waterfalls and from nearby a funicular leads to the **Dachsteineishöhle**, a cave of gigantic blocks of perpetual ice, and the Mammoth Cave.

Bad Ischl K8
Oberösterreich (pop. 14,000) Luxury shopping for a small town, for Bad Ischl remains a comfortable and well-patronized spa even though its imperial heyday as the unofficial court of Franz-Josef has long passed. Parks and woodland, riverside walks, and several buildings surviving from the grand days, especially the Emperor's villa (open June to September) with its royal domesticity. The Konditorei Zauner on Pfargasse is a famous left-over from the great days – the only truly grand Konditorei outside Vienna that preserves the legends of the famous Austrian confectionery shops.

Hallstatt

Linz F12
Oberösterreich (pop. 220,000) There are some fine Baroque buildings left in Linz, around the main square which is used today for concerts and entertainments as the trams clank past, but there are many better in Austrian towns with less distinguished names. Linz is no tourist trap, but its greatest days now are as a major industrial centre (steelworks and fertilizer, tobacco and pharmaceutical factories in the southern part of the town alongside the river), as a commercial centre, and the major port on the Danube. Industry is out of sight from the inner city but pays to keep alive the theatre, congress hall, and concert hall. Chief places to visit are: **the castle**, early 16th-century (closed Mon and Tues) with museum of Roman and pre-Roman finds and Gothic and Renaissance interiors; the little Carolingian church, the **Martinskirche**, near the castle, which is very moving; the **Landhaus**, former seat of the provincial assembly, with typical late 16th-century arcaded courtyard around a fountain – it was on the site of the Landhaus that Kepler taught and evolved his theory of planetary motion and Linz university, which is on the north bank, is named after Kepler. The Jesuit church (**Alter Dom**) is where Bruckner was organist for twelve years, the excuse, inevitably, for an annual Bruckner festival with the addition perhaps of Mozart's 'Linz' symphony (No. 36 in C, K425) or Beethoven's Eighth, both composed here.

A quick count reveals about 70 restaurants in Linz, together with coffee houses and bars. Centre of a rich agricultural region which fills the Linz markets. The drink in the surrounding countryside is cider rather than wine on which the rest of the Danube lives. Take a short outing by the very steep **Postlingbergbahn** (north bank) to the pilgrimage church on the hill for view of the city and the grotto railway; a longer outing (18km/11mi) to **St Florian**, birthplace of Bruckner, with its vast Abbey (regular tours April to October) with a magnificent library, picture gallery notable for works by Altdorfer, a warm, romantic contemporary of Dürer, and the Imperial apartments.

A macabre outing is to **Mauthausen**, down the Danube, where the granite quarries were used as a concentration camp and extermination centre by the Nazis and can now be visited with respect.

Lofer L3
Salzburg (pop. 2000) Lofer is an old market town, with huge block-like bourgeois houses, now mostly Gasthöfe, scattered higgledy-piggledy so that the sharp-cornered streets make this look like an alpine village. Minor summer and winter resort; outing to limestone caves and **Vorderkaser Gorge**.

Mittersill N2
Salzburg A plain little town surrounded by low hills, popular with hunters and the nearest town to the **Krimml** waterfalls, 30km/18mi west along a level valley. The total drop of the falls is 380m/1250ft making them the highest in Europe. The lower stage of the falls can be seen and heard from the road but the walk through the woods past the Gasthof to the mighty upper falls, say 3 hours there and back, is rewarding. The falls can also be approached from Mayrhofen and Zell am Ziller through the bleak Gerlos pass – the narrow side road through is more picturesque than the toll highway.

Radstadt N8
Salzburg (pop. 4000) A summer resort, enjoying high altitude in a gentle landscape of forests and country houses. The square, at right angles to the main road, is a parking street lined with low buildings in the Salzburg classical style; parts of the town wall and tower remain north of the main road.

Ried im Innkreis G7
Oberösterreich (pop. 12,000) The largest town of the justly unvisited part of Upper Austria called the Innviertel – a flattish, rolling landscape of fields and some woods, scattered with houses in a neo-tirolean style, with corrugated iron roofs. But in the cobbled market square of Ried (the Steltzhamerplatz) there are some fine old buildings, and people enjoying the sun, and some grander houses, green and red, in some of the other tucked-away squares. The other towns of the Innviertel are like that – quiet, unmemorable, friendly. **Mattsee** (Salzburg) is a pleasant little village with a nature park. **Braunau** is famous as the birthplace of A. Hitler (1889–1945; there is no monument). It has a rather distinguished old town, the outskirts are now industrialized. **Haag** rises from its tatty main square to the lovely **Starhemberg** castle and ski lift to the Luisenhöhe.

Salzburg J5
Salzburg (pop. 130,000) Salzburg, musical city of Mozart, beautiful city of the Baroque, lively city in a lovely setting. The throng of visitors, not only at festival time but for much of the year, adds to its happy feeling of enjoyment.

First, the festival. Mozart was born in Salzburg in 1756; a musical academy

Salzburg, from the castle

named after him – the Mozarteum – was founded about 100 years later; and in 1922 the first annual Mozart festival was held. Five weeks of music and drama until the last day of August. Presented in the modern Festival Hall, in the Mozarteum and other centres, its emphasis is on the divine master but with plenty of other composers and lyricists and always with the world premiere of an opera. The internationally famous perform there for an international audience. There are six or more performances a day to choose from, to satisfy most musical tastes, and the beginnings of a fringe festival as well. The city is packed out at festival time, and tickets for the more important events must be bought well in advance (through a travel agent.) For outdoor performances, some of the supporting programmes, and outside festival time, when the musical life is nearly as intense, you should get tickets on the spot at the official price.

Accommodation in Salzburg is generally expensive accommodation, and you may need to put up well outside the city if you are restricted to normal Austrian prices. The tourist office on Mozartplatz will help with this.

Even if you don't want to stay to listen to music, Salzburg is a lovely place to visit because the musical background and people create a happy atmosphere. The Old Town is squeezed between the left bank of the river Salzach and a 500m/1640ft high ridge called the Mönchsberg. On the right bank of the river is mostly the modern town, and it is easier to leave your car here in one of the multi-storey car parks and walk into the traffic-free old town. (There are two more car parks cut into the Mönchsberg at the back of the festival hall, but they get very full.)

Starting point for a walk round the old town must be **Mozartplatz**, with travel and tourist offices on one side, outdoor cafés at the end, and the 19th-century building of the provincial government on the other side. (If you've a liking for these things, even this is worth seeing, for the grandiose spaciousness and sunny corridors of power like a great world capital.) From Mozartplatz you walk through to the much larger **Residenzplatz**, where outdoor performances take place and on gala nights torchlight dancing. The **Residenz** which is one side of this square was built by the prince-archbishops who ruled Salzburg and beautified it up to the 19th century as their palace; a warm, pleasant, and magnificently stately home as a refuge from their gloomy fortress on the Mönchsberg. The palace is open to the public except when there are concerts or official receptions. Another side of Residenzplatz is formed by the **Dom**, cathedral, built between 1614 and 1655 by the most successful of the artistic archbishops, Paris Lodron. His predecessors had planned to build a Renaissance cathedral larger and more majestic than St Peter's in Rome; Lodron's mainly Baroque cathedral still has room for over 10,000 people, and makes the **Domplatz** which runs into Residenzplatz, a harmonious, superb whole. The cathedral square is used at festival time for

Northern Austria

Mozart's statue, birthplace and home

performances of the allegorical play *Everyman* by Hofmannsthal – the mystery of the rich man's death.

From a corner of the Residenzplatz opposite the cathedral one walks through to the **Alter Markt**, a small square lined with souvenir shops, cafés and downstairs cellar-bars, which gives access to the **Getreidegasse**. This little alley, always full of visitors, is two lines of high 17th- and 18th-century houses, one of which, No. 9, was the birthplace of Mozart. The ground floor is a wine shop, but the upper floors which are open to view contain mementoes of the composer including his violins and his spinet and are preserved much as when he lived his early life there. A pilgrimage that must move any music lover.

Parallel to the Getreidegasse is the **Universitätsplatz**, used today as an open-air market place. The old university building is no longer used for studies, but houses a fine smallish library of old books and manuscripts and next to it is the former university church, one of the finest Baroque churches in Austria. This, the **Kollegienkirche**, was designed inside and out by Fischer von Erlach. Behind this church are the festival halls – the large, over-imposing modern hall with room for 2300 spectators, and the smaller old hall, converted out of the original riding school and stables of the archbishop's court with superb ceiling paintings. These are right by the tunnel which leads through the Mönchsberg to a modern part of the town and on to the nearby airport.

Crowning the **Mönchsberg** is the fortress which you see from all over the city. You can get to it up stone steps which start from one or two little back alleys along the foot of the ridge (many of them still with little workshops for modern repairs in a setting Mozart would have recognized), but much more easily by the funicular, the station is at the top of a little street leading out of the **Kapitelplatz**, at the back of the cathedral. The fortress is called the **Hohensalzburg**; it was begun in the 11th century as a refuge for the archbishops in the wars between the Pope and the Holy Roman Empire, and its interior softened and made comfortable in the following centuries until the Residenz was built. The fortress is open to the public with guided tours (in English); best parts are the **museum**, with plans and prints showing the growth of Salzburg, the **Reck watchtower** with a splendid panorama of the Alps to the south, and the **Kuenburg** bastion with a splendid view of the bell-towers and domes of the city.

Apart from the prime sights mentioned above, old Salzburg is a mass of church buildings, a riot of Baroque architecture. Best of these is probably **St Peter's Abbey**, close by the funicular station; others are the **Nonnberg priory** by the Hohensalzburg, the **Franciscans' Church**, a 13th- century Romanesque and Gothic church near the Domplatz, and the **Holy Trinity Church** at the head of **Makartplatz** on the right bank of the Salzach, another Baroque product of von Erlach with frescoes like those ceiling paintings in the stables.

On the right bank, the chief things to see are the house where Mozart lived in Makartplatz, the formal **Mirabell** garden and remains of the **Mirabell Palace** (built by Archbishop Wolf-Dietrich for his mistress and twelve children, but largely burnt out in 1818; now used for concerts), with the great congress centre in the **Kurgarten** beyond and **Schwarzstrasse**, where are located the provincial theatre, the marionette theatre and the **Mozarteum**. The marionette theatre presents the classical light operas on record with beautifully carved wooden puppets on strings to play the parts.

Chief exhibit of the **Natural History Museum** is a mummified rhinoceros, a relic of the Ice Age discovered in 1929.

Most popular outing from Salzburg is to the palace of **Hellbrunn**, about 6km/4mi south. This was the summer residence of the archbishops with suitable banqueting rooms and octagonal music rooms, and lovely gardens with deer-park and fountains, some of them designed to spray innocent visitors from concealed jets! **Leopoldskron Palace** is within walking distance of Salzburg, a late 18th-century effort with beautiful grounds, now a pleasure park. The pilgrimage church at **Maria Plain** 5km/3mi north of Salzburg is exceptionally ornate even for the richest Baroque, and beyond Maria Plain is **Oberndorf**, another pilgrimage centre because the famous carol 'Holy Night' was written here on Christmas Eve 1818 (see Hallein, p. 121). For viewing points, the **Gasberg** (16km/10mi east) is favourite, with views to Salzburg and better to the Salzkammergut and Dachstein; another is **Untersberg**, 24km/13mi south on the road to Berchtesgarten, with a cable car to the peak at 1787m/5862ft.

The restaurants, the constantly changing nightclub scene, the casino at the Hotel Winckler, light entertainments, and Bavarian-ambience beer cellars and terraces, are the marks of a sophisticated internationally popular centre. There are large youth hostels and camping sites for the cost conscious.

Castle from Café Winkler: left; 'Salzburger Nockerl': right; Residenzplatz

St Gilgen

St Gilgen J7
Oberösterreich (pop. 3500) A quiet, rather charming resort on Lake Wolfgang, birthplace of Mozart's mother. Centre for boating (Union Yacht Club), surfing, good swimming beach (gravel) on Zinkenbach peninsula.

St Wolfgang J8
Salzburg (pop. 3000 – plus millions of visitors) A long village stretching along the lake, with its most famous hostelry the White Horse Inn overlooking the lake – there are many others, equally good, less crowded. The operetta *White Horse Inn* which, much more than the pilgrimage church with its masterly Gothic altar, makes St Wolfgang a modern centre of tourist pilgrimage, was written by Benatzky in 1927. Good outing by rack railway to the **Schafberg** for romantic views. Car parking horrendous – be prepared to walk some distance, from before the village (no room at the dead end beyond the village) or arrive by paddle steamer from St Gilgen or Strobl.

Steyr H13
Oberösterreich (pop. 40,000) The Steyr-Puch motorbike is built in Steyr, and BMW bikes and trucks are assembled here, but to push your way through the industrial rush hour to the peace and charm of the old town of Steyr is worthwhile. Its long main square, lined with houses from Baroque onwards in many different colours is one of the best-preserved street-squares in Austria. Many of the houses have handsome courts, like the **Heimathaus**, and the **Town Hall** and **Bummerlhaus** are noteworthy. Walk from the main square down to the river Enns through the '**Enge Gasse**'. Best view of the river front is from the bridge – up close it's a bit shabby.

Just outside Steyr is **Christkindl**, meaning Christ Child, a mauve-pink, Baroque church behind which is the Wirtschaft which at Christmas serves as a post office from which greetings can be sent all over the world and which answers all letters addressed (with reply paid envelope or stamp) to the Christ Child.

Tamsweg P10
Salzburg (pop. 5000) Tamsweg, coldest place in Austria, is the centre of a district called the Lungau, which is rather cut off and still strong in folklore – the ads in the main square are for performances of the local choir, bands, dancers, hunters' night processions. From the end of July and throughout August there are processions headed by the green-garbed giant about 8m/26ft high representing Samson. Some of the buildings in the main square are solid Salzburg-provincial, like the **Town Hall** of 1570; others are country peasants' houses looking out of place in a town, like the furniture store on the north side.

Mauterndorf, 11km/7mi away, is rather similar to Tamsweg; their Samson giant is red. **Moosham Castle**, between Tamsweg and Mauterndorf on a side road, is open all the year round except Mondays and has a folklore collection – scene of dreadful persecutions of witches in 18th century. Mauterndorf, at the foot of the Radstädter Tauern mountains, makes a stab at being a ski centre, but to get right into the mountains here you need **Obertauern**, which is just a ski centre, not a village, and very quiet in summer.

Wels G11
Oberösterreich (pop. 48,000) The modern town is industrial (chemicals) and also has a considerable agricultural market, but at its centre has an old town where time stands still. The **Stadtplatz** is a broad cobbled street, over ½km/550yd long, lined on both sides with grand houses of the 16th and 18th centuries – nothing modern or even 19th-century to intrude. Solidly built and richly decorated, showing well how an Imperial town looked before the motor car. Some of the ground floors have been opened up as shop fronts, more have been skilfully adapted to modern selling techniques; all the upper floors are untouched. The southeast corner of the old town, by the castle (open Wed, Fri, 1700–1900, Sat, Sun, 1000–1200) is even more authentic.

Zell am See N4
Salzburg (pop. 8000) A rather unattractive summer resort in a region of great scenic beauty – shops, big hotels, big buildings. Motor boats on Lake Zell, which gets quite warm in summer. Cable car to the **Schmittenhöhe** for panoramas and winter skiing; several other chair lifts; funicular to and from **Kaprun**, the foot of the Kitzsteinhorn for glacier and all-year-round skiing. Kaprun itself is the centre of a vast hydroelectric scheme for which the barrages fill the upper valley with a string of attractive lakes.

WESTERN AUSTRIA

For many visitors the received picture of Austria is the sound of its most alpine province, the Tirol. The cheerful accordion keeping the dancers in leather shorts at their work, the sharp crack of men in hunter's costume at rifle practice, jolly brass bands at endless festivals, the gentle knock of the carver's mallet, the bells on the cows as they move up to flower-bright pastures and the whisper of skis coming down the silent snow. These are not all of Austria, nor exclusive to Tirol, but put together, they typify Tirol, the largest part of our region of Western Austria. The province is wholly mountainous and nearly every settlement, from plush resorts to remote villages is organized to provide visitors with hospitality.

Tirol is not in fact uniform. The different parts, each with their own character, are: the Kufstein-Kitzbühel district in the east, where the sheer limestone crags of the Kaiser mountains and more rounded slatey Kitzbühel Alps enclose a district of busy timber villages with just the famous resort of Kitzbühel and the working town of Kufstein to lend a touch of sophistication. Further west, the lower valley of the river Inn is lined with small, old towns that grew up on minerals rather than alpine produce. Lake Achen on the north side of the Lower Inn is devoted to boating and steamer outings, while the Zillervalley on the south side remains the very haven of Tirolese gaiety – native love of life blending nicely with a cheerful way to earn a living.

The capital, Innsbruck, the bridge halfway along the river Inn, is a beautiful city in its own right, offering sophisticated food and entertainments almost within walking distance of still rural valleys like the ski-populated Stubaital and neglected Gschnitztal and Sellraintal. Much more rugged are the valleys west of Innsbruck that lead up to the Wildspitze, highest peak in the Tirol – the Ötztal and much less frequented Pitztal.

In the south west, the Upper Inn valley and the Paznaun valley are comparatively undeveloped for tourists. Here is a different sort of Tirol. The compact villages show their connection with the Romansch inhabitants across the Swiss border in the Engadine. To compensate, southwest Tirol includes the Arlberg district which can claim to be the ski-capital of the world, centred round the expensive resort of St Anton.

The north west is mainly the Lechtal Alps, most richly wooded, very peaceful but for the occasional avalanche, and still fairly remote and unvisited, with a strong wood-carving tradition; the more visited (and better equipped) parts of this district are the resorts like Lermoos and Ehrwald, within sight of the Zugspitze, and Reutte, the only real town of the district.

The Lech valley wanders through to the adjacent province of Vorarlberg, to the ski resorts of Lech and Zürs, which are on a par with St Anton in Arlberg for smooth sophistication.

The province of Vorarlberg is the smallest of all Austria's **Länder**. Its main parts are the Montafon valley, park-like in summer and growing as a winter skiing district, and the Bregenzerwald which is more of a wooded plateau, superb for walking. All have the famous Austrian warmth of hospitality, though in many ways are more similar to neighbouring Switzerland – the Montafon people retain family connections with the Engadine, while most of the Bregenzerwald was settled by those busy colonists who occupied the eastern Valais. The painted house style is rather like that of Appenzell, across the Rhine, which is even repeated in Tirol's Lechtal. When the Habsburg Empire collapsed, Vorarlberg voted to join Switzerland, but while the Swiss thought about it, the allies forbade it, and today the red-and-white flag of Austria's Babenbergs waves as proudly over Vorarlberg as anywhere else.

The little principality of Liechtenstein lies between Vorarlberg and Switzerland, on the Austrian side of the Rhine. It is in law a sovereign, independent state, the equal of Switzerland or the United States. In practice it is almost part of Switzerland; it has a customs and postal union with the Swiss Federal State and its only diplomatic representation is a legation in

Western Austria

Bern. Modern Switzerland and Liechtenstein form a sort of confederation, with the loose and theoretically breakable ties that were part of the old Swiss Confederation before Napoleon. But the ties of sentiment are probably more with Austria, to which Liechtenstein has been subject for most of its history. The people of the principality are even more attached to their independence, which permits tax and company law to make Liechtenstein a source of riches for foreign investors and resident nominees alike. The words of the Liechtenstein national anthem are less stirring than the tune, better known to British readers as 'God Save the Queen'.

Bezau D5

Vorarlberg The Bregenzerwald is the hilly, thickly wooded but essentially pastoral plateau that fills out the valley of the winding river Bregenzer Ache, and Bezau is its best-known village. The region is a hiker's paradise, but wood and cattle, not tourism are its main occupations, and in church and at band practice you may still see the traditional costumes of a hundred years ago – broad-brimmed hats on the men, golden, crown-like headdresses on the girls. Bezau itself, encircled by mountains and a mixture of nearby broadleaved trees with dense pines higher up, is a gentle, mainly modern village just off the river. Other villages of the region include: **Alberschwende**, where the steep climb up from Bregenz levels out into the plateau, busy with forestry products; **Egg**, ribbon-development in a sea of flowers – **Grossdorf** above Egg is prettier; **Mellau**, similar to Bezau; **Andelsbuch**, very large, very separate farmhouses surrounded by 'lawns' of cattle fodder; **Schoppernau**, at the foot of much steeper mountains, with a climbing school; **Schröcken**, in a sparser, more alpine setting, immediately before the steep climb to the Hochtannberg pass. Accommodation is fairly cheap in the Bregenzerwald; skiing facilities from most places along the valley are mainly for local fun.

Bludenz F4

Vorarlberg (pop. 2800) A very busy little town – textiles, chocolate factories, tourist excursion centre. In the town itself are some old gateways and medieval town houses, mixed up with modern shops and social centres. Barely 600m/1960ft above sea level, it is surrounded by mountains reaching 3000m/9800ft. Pedestrian precinct in town centre.

Bludenz is the gateway to three tourist valleys. The **Brandner** valley is a narrow precipitous gorge along its lower reach, and after its chief village, **Brand**, which is growing as a ski centre, there is a cable car to the austere, glacial Lünersee. The Great Walser valley takes its name from the people of the Valais in Switzerland who settled here and in Brand in the 14th century and whose mark can still be seen in the bright paint of the houses and the occasional local costume. Prettiest village of this valley is probably **Sonntag**, in a V-shaped alpine cleft; **Fontanella**, popular for skiing, is in a bleaker, higher part of the valley.

The third valley reached from Bludenz is the valley of the Ill, in the district called **Montafon**. This, an alpine park, is a string of little villages with onion-domed churches which are opening up more and more to tourism, for the valley has a summer mixture of alpine fertility and austerity and also good, unfashionable winter skiing. **Schruns** (market town) and **Gaschurn-Partenen** (ski centre) are the larger and more popular places. **St Anton**, **Bartholomaberg**, **Gortipohl** (concerts), **Gargellen** (lost at the end of its valley) and even the better-known **Tschagguns** (spa) are more intimate. At the upper end of the valley is the Silvretta range, which includes the artificial Silvretta Lake in its harsh, magnificent setting (seen from the dam in passing) and Piz Buin, 3132m/10,866 ft. The range is mainly for serious skiers and mountain climbers, and the road for experienced drivers – without a trailer. The lake is the source of extensive hydroelectric works, most visible down the valley at Vandans.

Bregenz C4

Vorarlberg (pop. 24,000) Almost two towns in one – the modern leisure centre spreads along the shore of the Bodensee (Lake Constance), with its Casino, boat cruises on Lake Constance (*eg* to **Mainau** for orchids in May, dahlias in September), riding, fishing and boating, some provincial shopping and the interesting provincial museum, carnival (revival of Allemanic tribal customs) at Maundy Thursday and Mardi Gras, and above all the festival. The festival, from mid-July to August, concentrates on light opera but includes theatre, more highbrow music, strolling players and minstrels, and Italian opera. The most spectacular performances are on a floating stage in the lake, watched from an amphitheatre on shore.

Above the festival town is the traffic-free old town, of large, old houses covered in red and gold creeper, silent and restful but a steep climb from the lakeside; look down on the roofs of the old town from the 13th-century **Martin's Tower**.

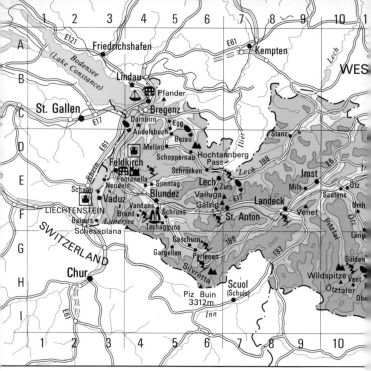

West of the town is a 4km/2½mi lakeside stretch of fishing harbour, yacht basins, sports and camping grounds, which ends where the Ache enters the lake. Here is the military exercise zone (alive with hares) and the newly developing nature park. In the town, about 400m/440yd from the main harbour, is the station for the cable car to the **Pfander**, a six-minute trip to a viewing point (1034m/3392ft) with a superb panorama, when the mist clears, along the lake to the Rhine, to the snows of Scesaplana, to the Bregenzerwald, to Germany, even to distant Glarus.

Dornbirn C4
Vorarlberg (pop. 40,000) Textile-manufacturing heart of the Vorarlberg, with a trade fair in mid-July to coincide with the festival in Bregenz 12km/7mi away. A sprawling town of factories and housing estates with a very small, but neat centre. **Hohenems Palace** is 7km/4mi south; cable car to the **Karren** (976m/3200ft viewing point) for a walk down to the Rappen Gorge or the smaller Alpoch Gorge.

Feldkirch E3
Vorarlberg (pop. 24,000) A very old town, its picturesque centre is the **Marktgasse**, a medieval rectangle bounded by arcades with oriels and painted façades, at its best when filled with tables and full glasses for one of the festivals that are a regular feature of Feldkirch life – especially wine festival in July. Visit the castle (**Schattenburg**) which houses a museum of aristocratic, bourgeois and peasant furniture. The new town, northeast of the old town centre, is the lively core of tourist life.

Hall in Tirol E14
Tirol (pop. 14,000) A former salt-mining town, as you can guess from the line of chemical works along the main road, Hall makes some attempt to be a spa under the name Solbad Hall, but its chief interest is in the compact, enclosed medieval residue. The centre of the old town is the **Oberer Stadtplatz**, still occasionally used as a market place, an irregular jumbly space between the parish church (close your ears at bell time) and the 16th-century Town Hall. Little cobbled alleys

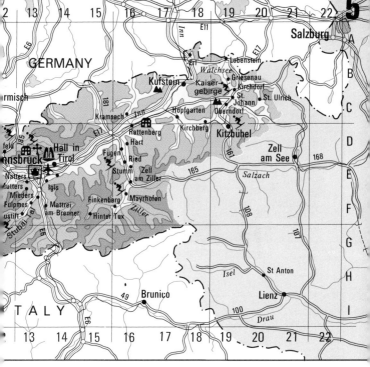

like the **Arbesgasse** and **Sparkassegasse** are lined by chequered old houses or oriel windows as in the Gasthaus Stach, while the **Rosengasse** leads to the small abbey square, severe 18th-century. The green-roofed tower is the former mint, closed in 1809 and reopened in 1975 to strike Olympic coins.

Imst E9

Tirol (pop. 6000) A market town just off a busy road junction, given variety by its sloping site so that the main square falls away from the high-steepled church. Base for outings into the **Pitztal** (less spectacular than many of the Tirol valleys, very restful) and up to the ski slopes by lift from **Hoch Imst**. Site of the first S.O.S. Children's Village. Every four years (or more? – the last was in February 1981) a masked procession called the **Schemenlaufen**, loud with the clash of cowbells on the dancers – it may be repeated next day, week or month impromptu for fun. **Mils**, about 7km/4mi up the Inn valley from Imst, offers swimming in the Inn.

Innsbruck E13

Tirol (pop. 150,000) Largest town in all the Alps and over 800 years old. From the main street, Maria Theresienstrasse, you can see the fierce slopes and sharp ridges of the mountains to the north, far enough away for Innsbruck to feel a city and provincial capital, but near enough for determined city workers to get in half an hour's skiing in their lunch break. Maria Theresienstrasse, loud with the clank of trams and the bustle of a rather elegant shopping street, contains some of the main sights of Innsbruck: the **Triumphal Arch**, built by Austria's favourite ruler, Empress Maria Theresia, to celebrate the marriage of her son and decorated by her as a memorial to the sudden death in 1765 of her husband, Emperor Franz I; the **Landhaus**, a Baroque palace dating from 1726 for the Tirolese provincial Diet; the **City Hall** (Rathaus) facing the column erected to St Anne to commemorate the retreat of the Bavarian army in 1703. It leads straight into the old town.

The old town, or inner city, is surrounded by a wall-like ring of buildings, following the line of the now-vanished

74 Western Austria

Little Golden Roof, Innsbruck

moat, pierced by arches which are the only way in. No traffic is allowed in the old town (after 1030) and in this pedestrian precinct are the chief sights of Innsbruck. Most famous is the **Goldenes Dachl** (little golden roof), built by the Emperor Maximilian as a Royal Box from which to view theatrical performances in the square below. The tiles of the roof are copper plated with gold – the story that they are pure gold tiles, put there by Duke Ferdinand the Penniless to refute the story that he was mean, is too good not to repeat just because it is not quite true. Performances by strolling players still go on in the square, but for the most part it is a place to sit in the early evening or at any time to take a cup and admire the Baroque and Gothic mixture of old buildings which make the inner city a delight. Many of them are rather pricey shops now, but see especially the **Helblinghaus**, 18th-century Rococo on the corner facing the Golden Roof, and the 15th-century inn called the **Goldener Adler**, with plaques engraved with the names of many of its famous visitors.

One wall of the inner city is formed by the **Hofburg** (open 0900–1600, closed Sunday), a palace built to the orders of Maria Theresia; its exterior along the road called Burggraben is severely classical in style, and in the ochre colour that is such a feature of her palace at Schönbrunn in Vienna, but the interior, sumptuous ornate state rooms glorifying the Habsburg monarchy, are a foretaste of the heavy imperialism of Vienna's Hofburg – see especially the **Giants' Hall** with its porcelain finish and painted ceiling, and the furniture and tapestries.

Opposite the Hofburg is the **Court Church** (Hofkirche), built in 1563 as the tomb for Maximilian of the Golden Roof (though Maximilian was actually buried in Wiener Neustadt) and containing 28 enormous statues of Maximilian's forerunners in majestic imagination, like Arthur of the Round Table and Dietrich von Bern (Theodoric the Goth, King of Verona), apparently designed by Dürer but the figures cast by workmen more used to casting cannon.

Next to the Court Church is the **Tirolese Museum of People's Art**, with a good display of house styles, furniture, costume and utensils from the vanished peasant past. Other museums in Innsbruck are the **Ferdinandeum** (fine art of the Tirol) and the **Tirolese Regional Museum** (mineralogy, old coaches, Tirolese Home Guard).

Other sights which are a pleasure to visit are: **Bergisel**, scene of several battles where the Tirolese under the national hero Andreas Hofer defeated Napoleon's French troops, now laid out as a park with woods and lawns; the Rococo **Basilica** of the suburb of Wilten and the Baroque **Abbey Church** of Wilten; **Ambras Castle**, like Bergisel on the south side of town, with art collections, armoury and lovely park (closed Tuesday); and on the north side of town the **alpine zoo**, the Panorama (Riesenrund-gemälde), a gigantic circular fresco depicting the battle of Bergisel, and the funicular up to the **Hungerburg** for a superb view of the valley and the Alps or for a continuation by cable car to **Seegrube** or **Hafelekar** for the skiing.

For fugitives from the slopes, Innsbruck offers concerts, operetta, brass bands, yodelling and zither, Weinstuben and restaurants in plenty but little late night life, shopping in Maria Theresienstrasse and adjoining streets and under the arcades of the inner city. Innsbruck Fair in September is also the occasion of the Tirolese Culinary Week.

Innsbruck is noted for its mountaineering course and tobogganing, but especially one of the great winter skiing centres. The 1964 and 1976 Winter Olympics were held here, with the great ski-jump, ice-stadium, speed skating track as their mementoes. Outlying villages like **Mutters**, **Axalm** and **Tulfes** are also used as a base if you want to be out on the first ski-hoist of the day; **Natters**, just north of Mutters, is more peaceful with its little lake; **Igls** is a fashionable alternative to Innsbruck itself – reached by tram from Bergisel, high in the hills, with woods and parks like a spa, very pleasant and correspondingly expensive.

Innsbruck

Kitzbühel   D19

Tirol (pop. 8000) One of the best-known winter ski resorts, with chair lifts, ski hoists and cable cars almost directly out of the town up to the runs of the 'Kitzbühel Skizirkus' which allows downhill skiing over an 80km/50mi circuit with the occasional mechanical hoist to regain height. Kitzbühel is still an old, alpine town despite the crush of mock-alpine and frankly modern accommodation round its edges. The centre is in the form of a street-square (closed to traffic most of the time) with substantial pastel-painted houses, now inns or shops, centred round the Gothic church. Great centre for competition skiing, slalom and especially ice hockey. Its lake, Schwarzsee, is a bit marshy but warm enough in summer for swimming. In fact the town was a popular summer resort, with outings to the relatively gentle and restful adjacent Alps and to the great viewing point of the **Horn**, before the ski industry became predominant.

Other places which offer access to the Kitzbühel Alps are: **St Ulrich**, very small between its lake and the shops; **Hopfgarten**, semi-industrial (wood and cement) with an estate of workers' houses but also a wide range of winter sports facilities because there are workers to use them; **Kirchberg** with a small bathing lake and a deer reserve; **Oberndorf**, small and scattered.

Kufstein   B18

Tirol (pop. 14,000) This ancient frontier town can detain you for half a day if you are content to visit its sights, or for a month if you like it as a base for mountaineering in the rugged limestone massif called the Kaisergebirge. The lower town with its promenade along the river Inn is a grey, early 18th-century collection of town square and squeezed-in street – usually very busy – and overlooked by a great rock crowned by a fortress, the **Festung**, dating from the 12th century (tours 0900–1200 and 1400–1700). This includes a museum, cellars and dungeons and the huge **Emperor's Tower** (Kaiserturm), once a prison. The **Citizens' Tower** (Burgerturm), at the entrance to the fortress, houses a giant organ which gives giant-sound recitals at midday (and at 1800 in summer). Two sites from which to get a good view of the fortress are **Pendling**, a mountain across the river, or the **Heroes' Hill** (Heldenhügel), with its memorial to Andreas Hofer.

The slopes of the Kaisergebirge rise abruptly from the thin strip of plain on which Kufstein stands, and there are ski lifts straight up into the mountains. The higher line of peaks is called the Wilder Kaiser, jagged crests, steep and snow-covered for most of the year, while the lower peaks in the north are the Zahmer Kaiser with some alpine pastures but still with summer snow at the summit and a challenge for mountaineers. The only real road into the Kaisergebirge is from **Griesenau** which is at the start of the **Kaiserbachtal** nature reserve. Resorts which give access to these mountains include **Walchsee**, which offers boating activities in summer, **Kirchdorf** which makes an effort at low cost accommodation in farms and private houses, and **St Johann** (p. 77). **Erl**, about 12km/7mi down the Inn valley from Kufstein, is the scene of a revived passion play once every five years (1983 is next) with the theatre used for religious concerts in other years. The ravine at Klobenstein, on the German frontier, is quite dramatic.

Landeck   F9

Tirol (pop. 8000) Spread along the valley as two parallel one-way streets of shops and restaurants, with steep stone steps leading up to the residential parts, Landeck is an industrial town, busy in daytime with its own affairs but reverts at night to the feel of a mountain village – Tirolean music and dancing in urban bars. Surrounded by forested hills, with cable car up to the **Venetberg** 2513m/8344ft, for more ski lifts; ruins of Castle Schrofenstein at **Stanz**.

Lech   E6

Vorarlberg (pop. 1400) A scattered resort of modern hotels, busy with amateur walkers in summer, very busy with professional-standard skiers in winter; competes with its neighbour and duplicate Zurs in attracting the strong-ankled and well-heeled international set.

Mayrhofen   F16

Tirol (pop. 3500) At the turn of this century Mayrhofen became known as the archetypal Tirolese mountain village, at the head of the Ziller valley, and even in 1945 it retained that atmosphere while depending heavily on trade as a summer-school centre allied to Innsbruck university. Today it is a small tourist town, of shops and street lighting, conference centre at the **Europahaus**, and package tours (very popular with the English); but its country base remains underneath – in May the cows mooch sedately through the village on their way to summer pastures, and the locals provide the round of Tirolean evenings, folklore shows and summer fetes that form the basic summer diet. Cable car to the **Penken**, for panorama, from top of the high street and to the **Ahorn** from the 'suburb' of Kumbichl. Plenty of winter skiing, with runs that lead right into the village. Access to the glaciers of the Hintertux valley for summer skiing. Trains on the little Zillertalbahn are occasionally pulled by steam engine for a fun outing.

The other villages of the Zillertal are less developed. **Zell am Ziller**, the largest, still has the feel of a village, with chalet-style houses grouped round the main square, though it is well geared for tourism. **Uderns**, **Hart**, **Stumm** and **Ried** are little hamlets in the broader part of the valley, basically pastoral but receiving visitors in the farmhouses that cluster thickly on the adjacent slopes, and with some skiing facilities. **Fügen** makes a play for activity holidays; **Finkenberg**, beyond Mayrhofen, is more hemmed in by mountains, to which it gives immediate access.

Rattenberg D16

Tirol (pop. 700) A medieval town which stopped development when the mines were exhausted in the early 1700s and has remained as it was then, compact and picturesque along the Inn. There is a castle just outside the town, and Achenrain castle surrounded by little lakes at **Kramsach** across the river.

St Anton am Arlberg F7

Tirol (pop. 2200) St Anton is essentially a ski centre, with nursery slopes nearby and runs for the expert from the Galzig and the Valluga peaks, but it's more than a collection of hotels, and at its heart is a mountain town. It was the birthplace of modern skiing technique as evolved at the turn of this century by Hannes Schneider and still one of the world's great centres for serious slope-work, though now very fashionable. Nearby – **St Christoph** – less smart. At nearly 1300m/4260ft the summer rambler is almost in hay-covered alpine pastures or up into the rocks.

St Johann in Tirol C19

Tirol (pop. 6000) A market town, with a little industry based on milk products and just off the heavy trucking route between Innsbruck and Salzburg, the heart of St Johann is large houses with painted murals and still larger inn in the old Tirolean style, grouped round the church. The sports and entertainments facilities are for the townspeople but enjoyed by visitors who use the town for access to the Kitzbühel slopes or the climbing in the Kaisergebirge without the cachet and expense of the big resorts.

Seefeld D12

Tirol (pop. 3000) Seefeld is a prominent health resort, with half a dozen high luxury hotels which may open only in the season and fifty or so less expensive open all year. Smart casino. It has also become very fashionable as a winter sports centre, partly because of the quality of the instruction and also because of its position, on a plateau between the Karwendel mountains in the east and the Wetterstein mountains in the west. These include the **Zugspitze**, highest peak in Germany, on the Austro-German border, which can be reached by cable car from Ehrwald on the Austrian side. **Ehrwald** is a very scattered, rustic village overlooked by the rocky treeless heights of the Zugspitze, and is reached by a car or train from Seefeld by a corridor through Germany. The adjacent sister-village of Ehrwald is **Lermoos**, more sophisticated; for a simple mountain village in this region visit **Berwang**, west of Lermoos.

Röss-hutte, Seefeld

Seefeld

In the Ötztal

Sölden G11

Tirol (pop. 1600) Largest of the resorts in the Ötz valley, whose lower part is a perfect picture of green or wooded alpine pastures, lovely for hikers, while the upper part beyond Sölden is more bare and rugged, home to skiers and mountaineers, leading up to the peaks of the Wildspitze (3772m/12,375ft) and the Ötztaler Alps. Down in the valley, **Sautens** which has modern-looking facilities and **Ötz** which is fairly undeveloped, are stopping places for little Lake Piburg, warm in summer and used for bathing and boating; **Umhausen** is the starting point for a hike to the **Stuiben Waterfalls**, a well shaded climb (take provisions – the Waldcafe by the bridge was closed) to a 150m/492ft high thundering cascade; **Längenfeld** is picturesquely angled round a steeply projecting mountain. Sölden itself is a rather straggly village, and its appendage **Hochsölden** is a scattering of sport hotels round the chair hoist station waiting for the early snows. Beyond Sölden, little unspoilt **Vent** at the end of its valley and **Obergurgl**, highest village in the Alps, with its higher appendage **Hochgurgl** give instant access to some of the highest peaks and marvellous views to dramatic crests which the glaciers are still scouring away. Obergurgl is a favourite hard-skiing centre.

Stubai Tal F13

Tirol A string of little villages south of Innsbruck which have now developed into busy resorts, in a calm, larch-clad valley that leads up to the **Hochstubai**, where there is all-year-round skiing in mountains over 3000m/9840ft, surrounded by glaciers. **Fulpmes** is the largest village in the valley (the village play about the French wars of 1809 draws visitors from Innsbruck); **Mieders** is compacted tightly round its church spire with three Lopen for (winter) cross-country skiing; **Neustift** on a green slope is a substantial stopping place. Inexpensive compared with some ski centres.

Liechtenstein: stamps and the castle above Vaduz

Vaduz F3

Liechtenstein Vaduz is the capital of the little state of Liechtenstein and its only real town. Most visitors come here just because it is the capital of a baby-sized Ruritanian principality, ruled by His Serene Highness the Prince von und zu Liechtenstein. They post a few letters with the princely stamp in the princely post office and move on. Vaduz and Liechtenstein are, in fact, very interesting, though not for the mere sight-seer.

The town is mainly two parallel streets meeting at their end to form an oval circuit, lined with refreshment and entertainment stops; very lively at night for Vaduz has money to spend. Away from the centre there is a rash of cosy villas and beyond, the equally cosy, clean unobtrusive factories of modern Vaduz.

The castle looms darkly over the town from its high cliff. It was put up in the 16th century as a grim, functional building and declined steadily for four hundred years until in 1905 the prince decided to live in his principality and made the castle habitable. It is used now to house a nearly endless stream of distinguished visitors and is not open to the public. A selection from the enormous private art collection in the castle is on view in the **Liechtenstein Museum** above the tourist office.

If Switzerland is a lesson to the world in how to choose the right industries for a small country, and make a success of them by diligence, Liechtenstein is a lesson to Switzerland. At one time its chief product seemed to be company records – many international firms chose Vaduz as their official place of residence because the country's taxes are so low, and there are more companies registered in Liechtenstein than there are citizens. That source of revenue can't last for ever and some years back they started cautious diversification into industry; an odd but carefully selected mixture of products – it is, for instance, the world's largest producer of false teeth. The factories are tucked away so that the main road from Vaduz to Buchs, which is the equivalent of California's Silicon Valley, still looks like a rural landscape.

The people reflect this – sober, industrious, hardworking, earnest. Vaduz has a certain night life and gaiety but decently restrained so that Zurich appears a vice den by comparison.

From Vaduz it is only a 3km/2mi walk down to the Rhine which forms Liechtenstein's western boundary. The flat valley is narrow here and richly farmed. A pleasant outing is up to the ridge of the mountain east of the Rhine to look across the valley. **Balzers** is a quiet tourist base for this part with views to the Swiss Alps. In the north the lowlands widen out in an area of wooded hills and tightly tidy farms.

Away from the Rhine and its lowlands Liechtenstein is mountainous, looking very like the adjacent Vorarlberg which is its eastern border. Villages to note – **Gäflei** where begins the Furstensteig (prince's climb), a walkers' path along the ridge overlooking the Rhine; **Malbun** which at 1600m/5250 ft has fairly reliable winter snow and with seven ski lifts is the centre of winter sports in Liechtenstein.

Liechtenstein has an area of 168skm/65sqmi, and about 25,000 residents are citizens of the principality. There is a 24-strong police force, a police reserve of 30, and an army of 1 – the prince being the general though the army was disbanded in 1868.

SOUTHERN SWITZERLAND

The Rhône and the upper Rhine have their sources within a few miles of each other and flow in opposite directions, one to Lake Geneva and the Mediterranean and the other to Bodensee and the North Sea, through a great rift which divides the central from the southern Alps. Our region of Southern Switzerland is mainly south of this rift. The three very different cantons, have sunshine in common – the very dry air of Valais, the Mediterranean warmth of Ticino, and the bright high-altitude radiation of Graubünden.

Canton Valais, originally French speaking, was partly colonized by Germans from Bern in the 13th century and grew up as the fief of a prince-bishop whose powers survived while Valais was an independent, unruly republic in alliance with the Swiss confederation. When the Valais joined the confederation in 1815, lawsuits and disputes over water replaced the fierce quarrels between French and German which had marked its history. The canton, outside the established resorts, remained determinedly backward until the 1960s. Canton Ticino, wholly Italian, was harshly governed as a subject territory of the old Swiss confederation, but when the new Switzerland, the Helvetic Republic, was formed under Napoleon Ticino hastened to join it. Agriculture remains very poor, but the sun brings riches from other lands. Graubünden is a Switzerland in miniature – formed from a number of tiny self-defence leagues, it joined the Swiss republic at the same time as Ticino in 1803. It still claims to be a republic in alliance with the Swiss, and is largely undeveloped.

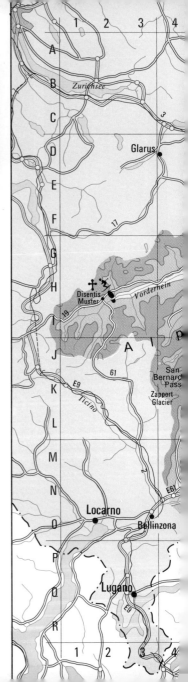

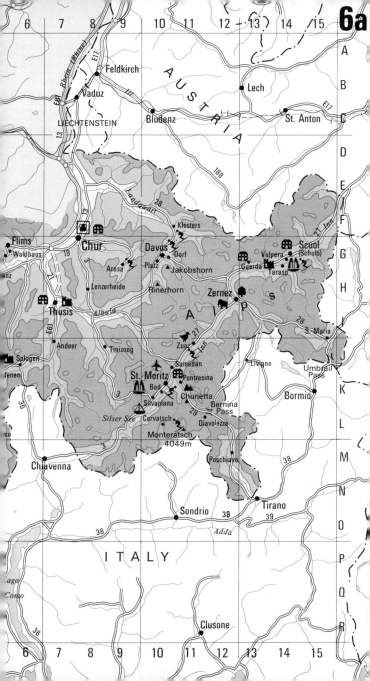

GRAUBÜNDEN

Largest of the Swiss cantons (7108sqkm/2745sqmi), one sixth of the area of Switzerland, with a population of only 170,000 – 3 per cent of the total – Graubünden is the least developed. (In Italian it is Grigioni, but is usually known by its French name Grisons, French not being spoken there.) It has remained a separate world within a world, a federation within the federation. There is a sprinkling of major ski resorts of world fame – St Moritz, Davos, Arosa – and plenty of other settlements ready for skiers and holidaymakers, but in comparison with Switzerland's other winter playgrounds Grisons is quite undeveloped – no rash of cable cars. For the most part this is Heidi country, pastoral meadows or virgin mountain. (Indeed, the Engadine publicizes itself as 'Heidi country', though the story is based on Maienfeld.)

The alpine scenery is rather different from elsewhere. The whole canton is high – average height 1500m/4920ft – but in place of the alternation of deep valleys with pastured mountain that characterizes most of the Alps, it is almost a high wooded plateau from which project the snow-covered peaks usually called 'Piz'. Because it is so high, the air is noticeably clearer and more stimulating than elsewhere, and this is why there are so many sanatoria in this canton. Also because of the height it is cooler at night in summer, even though in the daytime it shares the extra warmth of the other cantons of the southern region.

The chief river is the Inn, here called the En, and its valley, the Engadine, has come to mean winter sports for the wealthy. In fact most of the Engadine is untouched countryside or old villages in the highly characteristic building style with high towers reaching like the mountains for the clear sky. Most alpine villages, in Austria and in Switzerland, spread out generously, and only in the plains do the buildings huddle together, to save land; but in Graubünden the houses are all hunched up, as though to protect each other from the hostile world, and maybe this still expresses the traditional Engadine attitude.

The lower Engadine is marked by deep gorges and rushing waterfalls, the Upper Engadine has more lakes. The Rhine rises here. The Vorderrhein (front Rhine) starts in a deep cleft in line with the start of the Rhône, while the Hinterrhein (rear Rhine) starts from the Zapport glacier and cuts its way through deep gorges to join the main stream at Reichenau. The smaller Valser Rhein flows from a lake above Vals through steep-walled bare mountains and flowered meadows to join the main stream at Ilanz.

The traditional flora and fauna of the Alps – flowers, chamois, rare herbs, deer – are preserved best of all in the Swiss National Park, tucked away in a remote corner near Zernez. But the same can be found all over Graubünden, though it is getting less every year. Similarly the traditional and distinctive building style is preserved as official policy, for example at Zuoz and at Guarda. This cannot preserve a way of life; the old Engadine ways can be seen and if you explore the mountains you will find more of the forgotten past than anywhere else, but most of the canton is already up to date.

This is a conservative area – no cars were allowed here until 1927. Ten per cent of the roads are in tunnels. The highest pass in Austria or Switzerland (Umbrail pass, 2501m/8000ft) is in Graubünden. The railway, the narrow-gauge Rhaetian railway, is in places the most twisting, tunnelling, back-doubling piece of railway engineering in the world.

Conservatism is most active in the languages. One half of the people speak German, and about one sixth Italian, but the remaining third, say 60,000 people, cling to Romansch as their mother tongue. (Compare this with the 250,000 who speak Welsh.) Romansch is a direct descendant of the Latin spoken in these valleys when this was the Roman province of Rhaetia, a language in its own right parallel to French and Italian. It sounds like Portuguese spoken with a thick German accent.

The two main dialects of Romansch are Ladin, spoken in the Engadine, and Surveltisch, spoken in the Vorderrhein valley; there are local newspapers, books, histories, dictionaries for each dialect. Romansch is the fourth national language of Switzerland, but not a language of office, which means that it can be used in announcements of the cantonal authorities, and as the sole language in schools where the commune wants it, but outside Graubünden nobody is required to know it. This amounts to a sentence of death on Romansch, which accounts for the strenuous efforts which you may notice to encourage its use.

Before they joined the Swiss confederation in 1803, the people of Graubünden had organized themselves into three leagues, similar to the league of the forest cantons that was the start of Switzerland. The League of the Ten Jurisdictions was concentrated around Klosters; the League of God's House in the Engadine; and the League of Counts along the upper Rhine. The League of Counts was called in German the Grau Bund (from graw = graf = count), which was soon taken to mean the Grey League (from grau = gray) so it was called in French the Grisons. Graubünden or Grisons have now become the name of the whole canton.

The chief local handicrafts now are embroidery, which you can buy direct in Chur, and handwoven linens in Santa Maria. The local wine is Veltliner; red Veltliner is just a good wine, so-called green Veltliner is prized as something highly special, and rather difficult to get hold of. Much of it is grown in the Valtellina, now part of Italy between Poschiavo and Santa Maria.

For all that it is conservative, Graubünden is singularly short of local traditions to watch. There's the Plazidus festival in Disentis in July, and a country music rally in Arosa in August, and the Engadine music weeks (classical music) in July. The ancient custom called *Hom Strom* (Driving out winter) survives in Bad Scuol on first weekend in February.

Arosa G9
(ppp. 4500) A high altitude (1800m/5900ft) resort, tucked away off the main road, so it is fairly quiet as an internationally famous winter ski centre goes. Rather elegant without being flashy, surrounded by gentle woodlands of pines and larches where squirrels abound, and little lakes, some in the bare mountains, some in the green.

Chur G8
(pop. 33,500) Chief town of Graubünden (pronounced Koor), not of any great interest in itself but as a base from which to explore the canton, well equipped and with good communications to everywhere. The remains of the medieval town are between the 12th-century cathedral and the Grabenstrasse, and two foot-tours of the old parts are marked out for tourists with time to spend. The Rhaetian museum explains some of the complex history of this left-over league of leagues. There is a cable car to the sun terrace of Brambrüesch.

Lenzerheide is the nearest little mountain resort to Chur – a modern place in a north-south valley by a mournful lake, surrounded by park-like woods.

Davos G10
(pop. 12,300) A substantial business town in its own right as well as a famous ski centre, and also a minor spa (setting of Thomas Mann's *Magic Mountain*). Very popular for ice-skating, with the largest skating rink in Europe. The chief snow fields, shared with Klosters, are reached by the Parsenn funicular, and the Parsenn run is claimed to be the world's finest; there are many other skiing areas, like the Rinerhorn and Jakobshorn, and the Strela slopes for the less skilled.

Klosters is not quite another Davos – smaller, existing only as a resort, and more popular with younger skiers. Much prettier in summer, in an area that is quite rural but for the detritus of skiing. Huge swimming pool, children's ski-school.

Disentis H2
(pop. 3500) First town on the Vorderrhein, with an ancient Benedictine abbey (17th-century, on 8th-century foundation). The whole area has a cloistral quiet and peace, soft walks with the mountains not far off for winter skiing. A place to stop for the Romansch language and culture.

Flims G5
(pop. 2000) A family resort, with a small lake fed by warm springs to make it swimworthy in June, and an abundance of ski lifts from Flims-Dorf and from Laax just up the valley, while **Flims-Waldhaus** is scattered hotels hiding in the pine trees. The town for Flims is **Ilanz** (pop. 2000), which is much older, a market town for cattle, and former capital of the **Grau Bund**.

Poschiavo M12
(pop. 3600) A very Italian-looking mountain town in a fairly steep-sided valley, above a lake shut in between much steeper mountain slopes, surrounded by rocks and pines. Reached via the very attractive Bernina Pass from St Moritz, the road passing through an idyllic landscape of alternating larches, green valleys, and fierce corniche. From the Bernina Pass a side road leads through Italian territory to the Umbrail Pass and back into the Swiss National Park at Zernez; this road passes through **Livigno**, which is a customs-free zone and repays a visit if your petrol tank needs charging.

St Moritz K10
(pop. 6000) Perhaps the most famous fashionable ski centre of them all, and home of the Winter Olympics in 1928 and 1948, St Moritz is an unfashionable-looking place. There are two centres – **St Moritz-Dorf** is the town, laid out on a steep hill with shops and the super de-luxe hotels, but also with reasonably-priced accommodation for those who want to watch the rich, and from the town you look down on **St Moritz-Bad** in the

Southern Switzerland

St Moritz

valley, beside the lake, a spa based on iron-rich waters, used as a cure before Roman times but now an array of tower-block hotels looking like a French housing estate. The winter sports are superb, with the Cresta run for tobogganing, a high ski-jump, and all the ski lifts the jet-set could want. In summer, in the absence of czars and jet-setters, there is a determined attempt to offer high culture and low prices, which keeps St Moritz quite accessible.

Pontresina, about 10km/6mi from St Moritz, is more of a mountaineering centre, with outings into the Bernina massif for glacier climbs, especially from Diavolezza, a viewing point reached by cable car, and also from the Chunetta belvedere (views of Morteralsch glacier) which can be reached by road. The ski season here lasts most of the year; when the snows have gone, there are concerts under the pines.

Silvaplana, 6km/3mi from St Moritz, is on another lake and a sailing centre for Graubünden, with summer skiing high up at Corvatsch. **Samedan**, 7km/4mi from St Moritz, is a true Engadine village of narrow street and high houses, adjacent to the St Moritz airfield.

Scuol G14
(pop. 1900) Scuol is an old Engadine town, the old part of two paved squares lying just below the modern main road. It is the lively part of a three-part spa; **Tarasp** is the spa centre, in a flat valley; **Vulpera** is a group of hotels on a terraced hillside, alive with summer flowers now with a complex of Engadine-style holiday chalets. The pure white Tarasp Castle, gleaming above the trees on a pinnacle of rock, is home of the Prince of Hessen-Darmstadt, but usually open to visitors on summer afternoons. Essentially a summer centre, the triplet is opening up ski hoists and chair lifts for winter business.

Splügen J6
A mountain village of the Hinterrhein, controlling alpine passes, with a lake surrounded by rocky scree. From Splügen the road to the San Bernardino pass (which leads to Bellinzona) brings one to the Zapport glacier which is the source of the Rhine and then, over the pass or through the tunnel, to the hamlet of **Mesocco** on a level patch in the hillside, starting point for the Castle of Misox. The castle is not quite a ruin, the impressive massive remains of the greatest feudal fortification in Graubünden. The road through the Splügen pass into Italy leads to the **Pianazzo** waterfall, 250m/820ft high.

Thusis H7
(pop. 2000) An old Engadine town in a flat plain surrounded by dramatic mountain countryside. The nearest gorge is the Via Mala – the evil way – a narrow section of the Rhine where the old road, twisting and squeezing along the floor of the ravine, has become a tourist's goal (down 321 steps) now a new road has been built. The gorge is overlooked by the ruins of the Hohen Rhätien castle, Ten km/6mi south of Thusis, beyond Andeer, the **Roffla gorge** is a dramatic torrent beside a walkway which ends at the waterfall. The third gorge in the Thusis region is the Schyn defile, a deep ravine covered in dark green vegetation.

Zernez H12
This large village of square Engadine houses with low-pitched roofs and arched windows is the gateway to the Swiss National Park. The park is protected from all human activity – no shooting, tree-felling, animal-grazing, camping, flower-picking – except observing. There is nothing but the natural flora of edelweiss, alpine rose and valerian and larches, and the deer, chamois, marmots and occasional ibex. There are hilly paths through the 181sqkm/70sqmi of the park, and you may on no account leave these. But if you just stay in your car, you will see nothing but woods.

Guarda, 12km/7mi from Zernez (on the road to Scuol) is similarly preserved as a rustic 'museum village'.

Zuoz F11
Best-preserved of the Engadine villages; another starting point for tours of the National Park, and also the terminus of the Engadine marathon Ski Run.

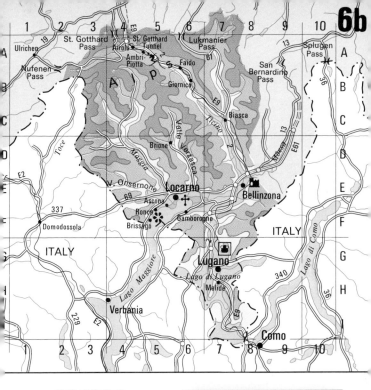

TICINO

Compared with the rest of Switzerland, the canton of Ticino seems Italian – in the Mediterranean warmth of its lakeside rivieras in the south, and in its stone-built houses in the steep valleys away from the lakes. It's Italian in language, with Italian faces and Italian hats, chestnuts and wired-up vines, Italian food with plenty of spaghetti. But if you cross the border into Italy you realize how very Swiss it is – clean, orderly, restrained with just a touch of southern gaiety.

Ticino was for centuries a captured subject territory of the original cantons, but when it became independent in 1803 it preferred to join the confederation as a full member. Today connections with the rest of Switzerland are mainly through the St Gotthard tunnels, but there are also connections across Italian territory by rail (to Brig via Domodossola) and by bus (Lugano to St Moritz); buses also travel to Ulrichen over the Nufenen pass, to Chur over the San Bernardino and to Disentis over the Lukmanier pass.

The north of the canton, the region of the St Gotthard or Leventina, is bleakly alpine – hard crests of mountain with no pasture, some remote villages still in the remote past, others with simple but not primitive hotels. The valleys further south are much softer though still in the same mould – in the wild Valle Verzasca, beginning at Tenero, Brione is a nice stone-roofed mountain village, while the Onsernone valley is a deep rift with more charming villages noted for straw-work. But they are not too backward – most have a little restaurant under the trees, called a grotto.

The most visited part of Ticino is of course the southern tip, around Lugano on its lake and Lake Maggiore with resorts like Locarno and Ascona. This is where the Swiss from other cantons come in pursuit of warmth for their holiday homes and the suburbs of separate villas round Locarno and Lugano are where they come for a holiday and stay to retire. These are the centres where you can enjoy a touch of Italian *dolce far niente*; palm trees, subtropical gardens, sunshine, luxury, and nothing to excess.

The red wine is a good Merlot and the white a rather inferior Nostrano, but there's also a rare red Nostrano which is worth finding. The food, while Italian, has some Ticinese specialities; the *ravioli*

Gandria, Lugano

Morcote, Lake Lugano

al pomodoro is usually quite special, *coppa* and *zumpone* are Ticinese sausages, the *zuppa al paese* is nearly always a thick vegetable soup, and *panettone* is a rich fruit cake.

Airolo · A4
(pop. 2000) A valley village which you must pass whether you arrive by car over the St Gotthard highway, or under the St Gotthard through the car or train tunnel. Most traffic from here charges south to the lakes along the main road which is steadily being up-rated, but if you've time take the old road which skirts the foot of the mountains through untouched villages. The bus route takes you to **Andermatt** over the St Gotthard and to **Ulrichen** in the Rhone valley over the bleak Nufenen. Take the cable car another 900m/2950ft up into the mountains for a long, lonely walk downhill to the train at, say, **Ambri-Piotta**. The surrounding mountains are hard, the rock bare or thinly green and gouged out to leave small lakes. Many waterfalls, some winter skiing.

Ascona · F5
(pop. 4800) A former fishing village on a flat delta on Lake Maggiore, now a quiet holiday resort, sports, sandy campsite, music festival Aug to Oct.

Bellinzona · E8
(pop. 18,200) Chief town of the Ticino – paved main street runs down from the station through a few rather elegant, very Italian, piazzas, lined with shops and restaurants. It has three castles: **Castle Uri**, in the town opposite the Migros; **Castle Schwyz**, up a steep ramp, fully restored; **Castle Unterwalden** higher up in chestnut groves, its café a a good viewing point. Trains meet here from **Lugano** and **Locarno**, trains also go direct to the duty-free zone of **Luino** in Italy, bus to **Chur** through the San Bernardino and Splugen passes. North of Bellinzona the valley of the Ticino closes in on the road to Airolo, densely planted with vines, southwards the valley opens out in a mixture of maize fields and industrial disfigurement.

Brissago · F5
A quiet, sunny village on Lake Maggiore, backed by steep hills, the lowest village in Switzerland. The tiny Brissago islands (botanical gardens) can be reached from here, or from **Ronco** or **Ascona**.

Faido · A6
(pop. 1500) A stop between the St Gotthard and Bellinzona in the middle of the attractively sparse Vallentina district; semicircular piazza, overlooked by sheer cliff, waterfalls in surrounding hills. **Giornico** is perhaps the most appealing village in this region on the main road, or **Chironico** if you want to divert.

Gambarogno · F5
A region of little lakeside red-roofed villages and hillside stone hamlets on Lake Maggiore opposite Locarno. Beaches, summer warmth.

Locarno · E6
(pop. 15,000) The chief resort on Lake Maggiore, hotels, gardens, sharply-curving lake-front promenade, from which narrow winding streets lead to the Piazza Grande where in summer the film festival is held. Very mild climate. Much quieter than Lugano. Funicular to church of **Madonna del Sasso**. Good centre for excursions into the nearby mountain valley (along the Centovalli to Santa Maria Maggiore is the favourite) and as far away as Venice. Steamer to Brissago islands, to Borromean islands and to Luino. Western part of the town, in the Maggia delta, is flat, laid out as a rectangular grid, the centre is curvy.

Lugano · G7
(pop. 29,000) The largest resort in the Ticino is very busy. It's on a steep hill by Lake Lugano, with a funicular between the station and the town centre. This, with a full range of fashionable shops, is a pedestrian precinct. Lakeside promenade with landing stages for steamers mixed up with boat hire is rather ornate, palm-treed. The Congress centre has ever-changing art exhibitions, jazz and classical concerts. In summer it is mild rather than hot, sunny, and can be humid; in winter it is mild. Good range of sports is available and an enormous range of excursions (*eg* funicular to Monte San Salvatore) for walks, cableway to Serpiano for woods and views, steamer to Melide for collection of Swiss houses in miniature, chair lift to Monte Lema for panoramas. Monte San Giorgio nature trail and fossils.

Ronco · F5
Don't go wrong Ronco, there are at least two, equally attractive but very different. Little Ronco All'Acqua is a hamlet in the mountains, bare, stone-built, on the road to the Nufenen pass. **Ronco Ascona** is a hillside Italianate village 150m/500ft above Lake Maggiore with palms and a campanile and a little harbour on the lake at **Porto Ronco** which is where the better hotels are. A third Ronco, on the old road from Airolo, you can pass through without noticing.

VALAIS

The river Rhône begins at Gletsch as a tiny trickle melting out of the glacier and rushes westwards towards Lac Léman, cutting a deep cleft through a sparse landscape of rocks and thinly grassed mountain slopes. Below Brig the valley widens as the river broadens out and is densely planted with vines or dead straight rows of fruit trees, or strawberry and asparagus. This upper valley of the Rhône (called the Rotten in the German-speaking part) is *the* valley, the spine of the canton of Valais.

North of the Rhône the central Alps, whose southern slopes are in the canton, keep out much of the cold, wet winds making this the driest and nearly the sunniest canton in Switzerland. There are a few resorts on the north side – the great complex of Crans/Montana, the little spa of Leukerbad, the lovely and almost undeveloped Lötschen valley, and mountain villages like Fiesch to give access to the mighty Aletsch glacier.

You can see the effect of this dryness in the plain, where irrigation pipes are constantly watering the vines; when the weather is misty elsewhere you are likely to get clear skies and bright sun. 'California of Switzerland' is an optimistic puff, but gives you the right idea. A string of sizeable towns, which have been working hard to industrialize since the 1960s, runs along the Rhône, but away from the towns every scrap of the flat plain is made productive.

South of the Rhône is the 'true Valais'. A dozen rivers rush down at right angles to the Rhône and have cut deep, steep, V-shaped valleys in the rock. As a result, the mountain villages are not, as in most of the Alps, scattered over an ice-scoured trough but are crammed along a narrow street

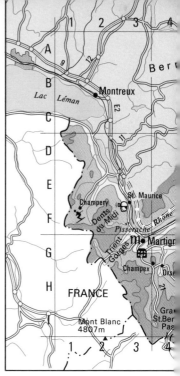

which runs along the mountainside. The houses are wooden, but not in the usual 'Swiss chalet' style. Instead they are four and five storeys tall, dark brown and undecorated. Next to the houses is often a wooden store-house supported on wooden pillars each capped by a broad overhanging stone to keep out the rats. These are very distinctive villages, and have clung tightly to their old traditions. The old costumes are still worn by a few women as everyday wear, and by many on fete days. When the cows are brought down from the alpine pastures at the onset of winter there are still the age-old processions and 'cow-fight' to find the queen of the cows – and not merely an old ceremony preserved for tourists. The French-speaking part of the Valais, roughly westwards of Sierre, preserves the past most tenaciously; most of the villages offer some skiing but have not turned into resorts. In the German-speaking east Valais are some of the big name resorts that have grown up because the slopes or the climbs are exceptionally good, but there are a few places that have withstood the appeal of visitors' money.

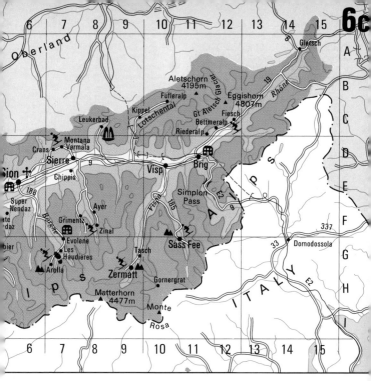

The characteristic dish offered in most restaurants is the 'assiette valaisianne' (or more rarely Walserteller, which is the same thing), a plate of mixed cold meats; *raclette* (p. 24) is the Valais version of fondue. The white wines are Fendant and Johanissberg, while the red wine, Dole, from the upper Rhône, tastes very like a French Côtes du Rhône.

Arolla   H7
Marmots and mountain goats roam around Arolla, an undeveloped little village high above Evolene, typical of many that you can find if you follow your nose. In summer, wild flowers carpet the meadows, in winter there is a little skiing. As a stopping place on the **Haute Route**, a long distance ski trail from Chamonix or Verbier to Saas Fee, Arolla is a centre of alpinism or less ambitious mountain exploration.

Brig   D11
(pop. 9500) The curious onion-topped domes and Italianate arcades of the palatial Stockalperschloss are an eccentric memorial to the 17th-century tycoon Gaspard Stockalper who put Brig on the map and made a fortune developing alpine routes. He became too rich for comfort and was chased out by jealous townsfolk. Brig is now a busy rail junction where the broad gauge from Geneva changes to the narrow red train that crawls through the Furka pass to east Switzerland. Outside Brig are the twin Simplon tunnels running over 8km/5mi through the mountains to Italy. Instead of the tunnel, sightseeing motorists may prefer the all-year pass through gorges, tunnels and snow galleries. There are two hospices en route – one run by monks the other built by Stockalper. If the pass *is* blocked, put your çar on the train at Brig.

Champery   F2
Ski centre of the French Valais, in the shadow of the Dents du Midi, with 161 ski lifts and 480km/300mi of pistes. Summer holidaymakers can swim, golf, tennis, walk in the forest or sit in the sun listening to the village band.

Champex   H4
Elegant without being fashionable,

Champex is a ski centre by a small boating lake with a beach, surrounded by dense woods. Said to be the most beautiful spot in the Valais Alps at 1500m/5000ft high, it has 1500 beds, and a superb view towards Mont Blanc. In spring and early summer it becomes an alpine garden.

Crans/Montana   D7
The most sophisticated of the ski-cities for high fliers, with the finest crop of high towers in the Alps. Three five-star and twelve four-star hotels. It looks north towards the fierce mountains and south towards the sun. **Montana** is the less posh part of the complex, on a patch of level, chalet-strewn ground like a mountain park. Viewing points are the **Bella Lui** to see the Alps of the Vaud, **Vermala** to look across the Rhône to the Anniviers valley with the Matterhorn at its head, and the **Plans Mayens** for a distant view of Mont Blanc.

Evolene   G7
Local folk in national costume mingle with holidaymakers in the narrow streets of Evolene village. In the brown larch chalets, high-rise in wood, you can still see craftsmen working as they have done for centuries. Further up the valley is the similar but smaller hamlet of **Les Haudières**, clinging to the mountainside above the steep drop to the river Borgne.

Fiesch   C12
(pop. 700) A halt on the Oberalp railway, with scattered chalets high on the Alps above the station. From here a Seilbahn flies up to **Eggishorn** for excursions to the **Great Aletsch glacier**. **Riederalp** and **Bettmeralp** are adjacent small resorts for skiing and the glacier.

Gletsch   C12
A handful of grey stone buildings in a grey landscape, at the junction of three passes, Gletsch is close by the Rhône glacier. The thick blue ice has been hacked out to leave a grotto where you can be photographed with guides in polar bear outfits.

Leukerbad   C8
(pop. 1200) A spa whose springs, at 41° are the hottest in Switzerland. Many long range ski pistes to attract a winter crowd.

Martigny   G3
(pop. 11,000) Square-streeted, square-housed, is this very square French town surrounded by vineyards. A Gallo-Roman settlement has been excavated at Octodurum and the collection of treasures from this is outstanding. Modern life is based on the huge factory for monumental glass. Nearby are the **Trient gorges** and the **Pisserache** waterfall. This is the start of the road to Aosta over the Great St Bernard Pass. The famous grey hospice, staffed by the priest-guides and their St Bernard dogs, is still there for the needy. A tunnel has taken the terror from the pass, and straightforward tourists no longer get free accommodation in the hospice.

St Maurice   E3
(pop. 3500) The old main street of this little town in the Rhône valley is narrow and brick paved, lined with little low shops, while the new road curves round the outskirts, lined with modern, gleaming white villas. The Grotte aux Fees is a subterranean gallery with lake and waterfall. It is here that the Rhône begins its metamorphosis from a tumbling mountain torrent to a broad flowing river.

Saas Fee   G11
A very 'smooth' ski resort on a gentle slope overlooking the steep fall to the Saas river. Cars are not allowed in the village and must be parked in one of the many car parks round the edge. As a mountaineering centre and terminus of the **Haute Route**, the long-distance ski trail from Chamonix or Verbier, it offers all-year-round sport on the higher slopes, and alpine walks from the village in summer.

Sierre   D7
(pop. 11,000) Reputedly the sunniest of Swiss cities, Sierre lies on the curious site of a prehistoric landslip. The modern market town sells fruit, vegetables, strawberries, asparagus, and a very good wine, though its prosperity comes from the aluminium smelter at Chippis.

Sion   E6
(pop. 22,000) As an important bishopric, once ruler of the Valais, Sion's ancient streets echoed with ecclesiastical pageantry for 1500 years. Today it is a quiet modern town with the old quarter hugging the cathedral. The 17th-century **town hall** with its ancient clock, the ostentatious **Maison Supersaxo**, the 16th-century town house of Georg Supersaxo, and the **Musée de la Majorie** are all to be visited. Two hills dominate the town – the hill of Valère houses the battlemented church of **Notre Dame** and the hill of **Tourbillon** is crowned by a bishop's fortress. There are guided tours, and a music festival from July to the end of August.

Verbier   G5
A growing winter ski resort, with new

Zermatt and the Matterhorn

chalets and some apartment blocks spreading fast over the floor of the Bagnes valley; popular for package tours, prices under some restraint. The pre-ski village survives, but for how long. **Haute Nendaz** and **Super Nendaz** nearby are rather more de luxe. This is one of the starting points of the Haute Route.

Zermatt   H9

(pop. 3000) Vies with Saas Fee for ski chic. Only a small place, but it has the largest area of summer skiing in the Alps. Very popular centre for mountaineers because the sharp peak of the Matterhorn and nearly 50 similar peaks are reached from here. There is a cogwheel railway to Gornergrät for a sweeping view over the glaciers, or to Rothorn for even more fabulous mountain panoramas. Cars must be left in Tarsch, a small village 5km/3mi outside Zermatt.

Zinal   F8

At the head of the Val' d'Anniviers, this is a developing ski resort; **Griments** 5km/3mi down the valley and 100m/330ft lower is an old village of traditional narrow streets and wooden houses, but it too is beginning to sprout ski lifts.

HEARTLAND SWITZERLAND

This alpine core is the part of the country that means Switzerland to visitors who don't appreciate how diverse Switzerland really is. One half is the Berner Oberland (the 'bare-knees' highlands) with the glacier-backed high mountains round the famous Jungfrau; much of it too high for alpine pastures, the territory of mountaineers and skiers, but attractive for its sheer majesty. The other half is based on the lake that ends at Luzern, the Vierwaldstättersee, the very picture of a romantic lake with fjord-like arms cutting through forested mountain. This region symbolizes Switzerland to the Swiss themselves, for it was here that their confederation was founded in 1291 when the men of the three forested districts, Unterwalden, Uri and Schwyz, which became the first cantons, swore to stay neutral and defend each other in the coming Habsburg wars. (These cantons, along with Luzern, give their name to this lake which they surround.)

The hero of that era was William Tell who was arrested for disrespect to the Habsburgs, ordered to shoot the famous apple from his son's head, and imprisoned for keeping back a second arrow to use against the bailiff. During the boat journey to the dungeon Tell escaped but returned to kill the bailiff at Küsnacht, starting a wave of revolt against Habsburg tyranny. That Tell existed is improbable, and even the revolts are doubtful, but the legend has persisted and grown so that the image of the indomitable fighters for freedom was the Swiss picture of themselves until quite recently. The play, *William Tell*, is still enacted every summer (at Interlaken and Altdorf) and there is no shortage of Tell souvenirs round here, though modern Switzerland has rather gone off this romanticizing of the past and prefers to live on its current achievements.

It would be hard to find a sturdy peasant fighter round here today, for the region has been occupied by tourism for a good hundred years. There is a wealth of cable cars and mountain railways to take you to many of the peaks, but tracts of country have escaped conversion into a pleasure park. Most villages that began life as somewhere for the peasants to live and work have now become resorts, living mainly off visitors, but they have remained small villages, sprawling a bit over the valley perhaps but still with a friendly, small-scale atmosphere. Of the towns,

Luzern, thronged with visitors, retains a sweet life of its own, and busy Thun is relatively unvisited; Interlaken was created just for hotels, but with good reason in its surroundings.

Some parts of the region are just an updated version of the old Swiss heartland – the lower Simmen valley is mainly devoted to feeding the flecked brown cows, and the canton of Glarus is a curious mixture of mountain industrialization and farming sewn together by a reduced tourism.

The Berner Oberland is the great area for sportsmen, being high enough to have reliable snow for skiing in the winter months. The Luzern area – the Swiss tourist office's region of Central Switzerland – is more for scenery and sightseeing, for the snow season is, apart from a handful of places, fairly short and unreliable. Where there is snow, it gets packed at weekends with Swiss who can take impromptu advantage of it, rather than with foreigners who have to book ahead.

Andermatt   L15
Uri (pop. 1600) A compressed, pretty little town of one wooden street in a broad, sombre valley, crossroads of the high Alps (altitude 1450m/4760ft). At Andermatt the toiling north-south road which crosses the Alps at the St Gotthard pass, meets the east-west cleft of the Rhine-Rhône valley. Below Andermatt the Gotthard rail tunnel leads to the sunny south from Göschenen, 20km/12mi away. The world's longest road tunnel, the 16km/10mi St Gotthard Autobahn, disappears from daylight here. A little red train shuttles back and forwards between Andermatt and Göschenen through the Devil's Bridge gorge. Andermatt is most alive in winter, when it appeals particularly for Langlauf skiing up the valley

Heartland Switzerland 93

to Realp; **Göschenen** is more a base for alpine hiking, *eg* to the crystal cavern at **Sandbalm**. The 13th-century church of St Columban in Andermatt traces its history back to 712.

Brienz J10
Bern (pop. 3000) This little lakeside village (571m/1870ft), spread along the hilly road that runs by the north shore of Lake Brienz, is a great wood-carving centre, and supplies many of the locally made animals and solid pictures that you find in souvenir shops all over the Berner Oberland. There is also a tradition of violin making. From Brienz you can hear the roar of the **Giessbach** falls, on the opposite side of the lake. From the village they look a mere trickle but a boat trip followed by a ride on the funicular brings you right up to a most impressive set of cascades. A little rack-and-pinion train pulled by a steam engine takes you, in 55 minutes, to the **Brienzer Rothorn**, which gives a tremendous panorama across the lake to the high Alps. Just outside Brienz is the Open Air Museum of Rural Building and Homelife.

Brienzer-Rothorn cog railway

Einsiedeln F16
Schwyz (pop. 9600) Here is the finest Baroque abbey in Switzerland, making Einsiedeln one of the great pilgrimage centres of Europe. Built in 1735, dominating the town, with a graceful exterior to make its ornate interior seem still more exaggerated, the church houses the Black Madonna which could be a thousand years old. Every year, on 14 September there is a torchlight procession to mark the Feast of the Miraculous Dedication, which recalls that when the bishop was about to consecrate the first church on this site, a ghost called out that God had already performed the dedication.

In front of the church, the huge town square is used one year in five for performances of a play by Calderón, *Great Theatre of the World*, a drama with all human life therein. In the intervening years, the monks coach the townspeople for the next performance. Einsiedeln is also a standard summer-outing, winter-skiing resort, and the birthplace of Paracelsus.

Engelberg I13
Obwalden (pop. 3400) The nearest resort to Mt Titlis and the largest winter-sports resort of the Luzern region, Engelberg has remained fairly unsophisticated and inexpensive. It is set in a broad, flat valley completely surrounded by mountains, and the buildings are mostly honest, squarish, brick with fewer wooden chalets than elsewhere in the mountains. The huge Benedictine abbey is now a religious college; the church has a splendid organ.

Frutigen L7
Bern A scattered village of low houses looking out to the intersection of three valleys; rushing streams for canoeing from the campsite. Here are the ruins of the Tellenburg, whose picture by the side of the railway viaduct you must have seen on publicity material. **Adelboden**, higher up the valley from Frutigen, is more compact and more of a ski centre.

Glarus G19
Glarus (pop. 6200) Chief town of the little canton of Glarus, which is a blend of industry and mountain-tourism. The town itself, although it has a rather impressive position in a ravine, is plain and regular, having been rebuilt after the great fire of 1861. Its chief interest is on the first Sunday in May, when the entire population meets in an open-air 'parliament' at which each member represents only himself (or herself in these equitable days) to speak and vote on cantonal laws and taxes. (See Appenzell, p. 103.)

Despite the textile and electrical industry along the valley of the Linth there are several mountain resorts in the canton, which includes the Todi massif (3620m/11,860ft) where the first hut of the Swiss Alpine Club was built. **Linthal**, reached by train from Glarus or, more impressively, by bus over the Klausen pass from Luzern, really goes for low prices; **Braunwald** (no cars allowed; arrive by funicular in 8 minutes from Linthal) is higher in the mountains; the Sernftal has the usual run of cable cars, ski slopes, toboggan runs, night skiing and cross-country tracks, centred on **Elm**, and in summer you can follow the route of Suvaroff retreating before Napoleon.

Interlaken K8
Bern (pop. 13,000) Interlaken, between Lake Thun and Lake Brienz, is the starting point for the classic alpine resorts of the Jungfrau region, and fun city for the

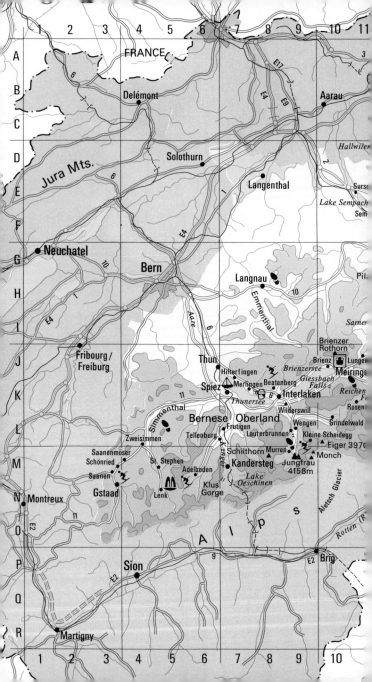

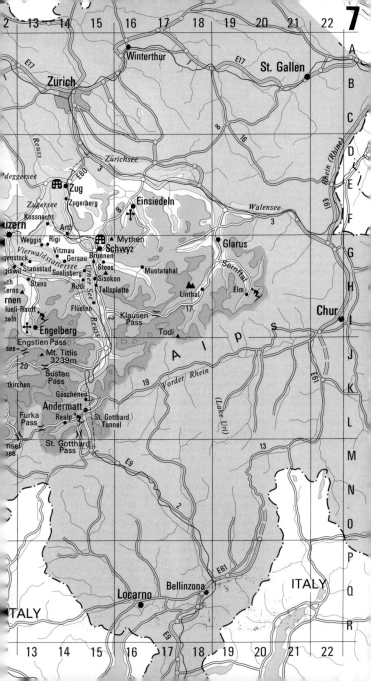

weary fugitives from the high Alps. It exists solely as a tourist capital. There are two stations, the east and the west station, connected by the famous promenade called the Hoheweg where souvenir shops outnumber even the many hotels. The eastern part of the town, and the suburb of Matten, is extraordinarily quiet – expensive detached villas half hidden in their plush grounds, each an island – but includes the meadow-theatre where in summer Schiller's version of *Wilhelm Tell* is played.

The entertainments and shops, the neo-Baroque casino and the neo-brutalist hotels are concentrated around the west station, and in summer the town is thronged. The general impression is of an opulent reserve for the idle rich, but there's room and welcome for the idling hard-up too, and the views to the Alps are magnificent.

Interlaken is a great centre for excursions – the chief ones near to foot are: to the **Harder Kulm** for the Harder Ibex reserve and viewing point overlooking Lake Brienz and the Jungfrau; to **Schynige Platte**, a still more noted viewing point (2100m/6890ft) with an alpine botanical garden reached by funicular from Wilderswil; and especially to the dramatic heights of the Jungfrau region.

Jungfrau K8

Bern The Jungfrau at 4158m/13,641ft is the highest of the three peaks (the others are the Mönch, 4099m/13,448ft and the Eiger, 3970m/13,024ft) which you see from Interlaken and which make the Jungfrau region and the mountain resorts on its north side the very symbol of alpine mountaineering and skiing. Most famous resort of the region is **Grindelwald** (alt 1034m/3320ft) with its one main street and scattering of chalets over a wide valley, not quite shut in by the majestic bare mountains. The gorge from which the Lower Grindelwald Glacier issues is

Staubbach waterfall, Lauterbrunnen

Stechelberg from Mürren

about 5km/3mi from the village. Grindelwald is the only resort in the gentle valley of the Black Lütschine, but there are several reached along the steep-sided White Lütschine. **Lauterbrunnen** (alt 800m/2620ft) is the largest, and is essentially a summer resort, in a steep-sided gorge over which cascades the Staubbach waterfall. Make a short outing to the Trümmelbach falls, most impressive of them all. By train from Lauterbrunnen, **Mürren** (1650m/5341ft, no cars) is another favourite rambling centre and **Wengen** (1276m/4114ft, no cars) is a skiing centre. **Kleine Scheidegg** (2061m/6762ft, no cars) is reached by train from Lauterbrunnen or Grindelwald, and is a resort for serious skiers, at the foot of the Eiger glacier. Two dramatic outings by cable car are to Stat Mannlichen (from Wengen or Grindelwald) at 2230m/7316ft and to the famous Schilthorn (from Mürren) at 2970m/9744ft, topped by the revolving restaurant featured in the Bond film *On Her Majesty's Secret Service*. This is the start of the world's longest downhill race, the Inferno, ending at Lauterbrunnen.

Highest outing of them all is to the **Jungfraujoch**, the yoke 3454m/11,331ft high between the peaks of the Jungfrau and the Mönch, reached by a rack-and-pinion railway that tunnels through the Eiger from Kleine Scheidegg. This is not, like the Schilthorn, a summit from which you look across to the mountains but the highest railway point in Europe, from which you still look up to the dominating peaks that only mountaineers can reach. The Jungfraujoch is popular for summer skiing, up in the land of perpetual snow.

Beyond the line of the Jungfrau – Mönch – Eiger are the icy wastes that separate central from southern Switzerland, especially the great Aletsch glacier. Once seen only by the brave, they are visible now by helicopter. No road crosses this vast white waste.

Jungfrau from Wengen

Kandersteg M7

Bern (pop. 1000) Chief appeal of Kandersteg, for non-skiers, is Lake Oeschinen, reached by chair-lift and an hour's walk, and probably the prettiest of the small alpine lakes, ringed by cliffs and overhung by snow-bound peaks of the Blümlisalp. Four kilometres (3mi) away the Klus is a wild gorge with the rushing falls of the Kander river. Kandersteg itself, where the Lötschberg rail tunnel begins (cars carried 14.6km/9mi to Goppenstein and on to Brig), is on a wide plateau and while this has allowed the village to sprawl and scatter, it affords plenty of fairly level walking in the immediate vicinity.

Langnau H8

Bern This market town is the cheese-centre of the Emmental, home of the famous Emmental cheese. The valley is a lush, green pastoral idyll, within sight of the high Alps, but itself just rolling hills with isolated farms and the cows. Good, undramatic hiking area with some phenomenally tall pine trees.

Blue Lake, near Kandersteg

Luzern F12

Luzern (pop. 68,000) Luzern feels remarkably gay (in the sense of cheerful) as Swiss towns go. It began life as a fishing village at the head of a lake whose beauties meant nothing to the shore dwellers, and remained just a small market centre for the original four forest states that were the nucleus of Switzerland, while the other Swiss towns were prospering as European cities. When the cities turned Protestant, rural Luzern remained Catholic. Luzern became internationally known in the 19th century, when the railway brought romantic tourists, hungry for the beauties of its many-armed lake fringed by cliffs and meadows and pine-clad mountains. Today it is the base for excursions to the sights and seclusion around the lake and a lively centre for holidaymakers.

The busy part of the town is very compact. The station faces directly the landing stage for lake steamers, where the lake narrows to become the river Reuss, and from here you cross the river to the old town by the wood-roofed footbridge (Kapellbrücke) decorated with 120 wooden paintings (early 16th-century) of the history of the town. In the old town there are several pretty squares – see particularly the 17th-century **Town Hall** and the wine market ' – and twisting medieval back streets with modern shops and a permanent throng of visitors. Fish market on Friday, general market Tuesday and Saturday, vegetable market along the quays every day.

Beyond the old town, starting from the gardened and casinoed quay on the north shore of the bay, is the more modern part of town with the more chic shops. From here the Löwenstrasse leads to the panorama (a huge painting showing the retreat of Napoleon III's army in 1871), the Lion memorial (much photographed stone carving of a dying lion, in memory of the Swiss troops who died at the start of the French revolution, which moved even Mark Twain to seriousness), the Alpineum (painted panoramas of the Alps) and the Glacier garden (32 potholes up to 9m/30ft deep, with a museum of prehistory and old Luzern). The museum of transport and communications, with early steamboat and first rack-railway, includes an information centre and planetarium and keeps children busy on a wet day. The art museum between the station and the

Luzern

lake is home to concerts during the Luzern Music Festival in August; folk festivals galore; carnival in winter; summer night festival with fireworks over the lake.

At least one of the three nearby mountain belvederes is a must for visitors to Luzern – the fairly low but strikingly beautiful Burgenstock (128m/420ft), the austere Pilatus, highest of them all (2132m/6995ft) or the more homely Rigi (1797m/5824ft). The Burgenstock, a mountain ridge projecting into the lake, can be reached by car or by funicular from Kehrsiten ($\frac{1}{2}$ hour by boat from Luzern); hotels and golf course at the top. Queen Victoria went up Pilatus on a mule, but it's easier now to go up either by the world's steepest cogwheel railway from **Alpnachstad** (reached by train or steamer from Luzern) or by a succession of cable-cars from **Kriens** (trolleybus from Luzern). You can go up one way and down the other. On foot it's said to take 5 hours (from **Hergiswil**). If you're keen on the glories of the mountain top at sunrise, then choose Rigi rather than Pilatus – there's less chance of mist. The peak can be reached by rack-railway from **Vitznau** (one hour by steamer from Luzern) or from **Arth** ($\frac{1}{2}$ hour by train from Luzern).

The correct name of the lake is Vier-waldstättersee – Lake of the Four Forest Cantons, the original Switzerland, but is usually called Lake Lucerne in English.

Many of the pleasure boats on the lake are still paddle steamers, and they all have a restaurant or bar on board. Regular services connect Luzern with all the lake-side villages and a short trip on the lake is as mandatory as a cable car to one of the belvederes, but the best part of the lake is at the end furthest away from Luzern, the blue fjord called the Urnersee – Lake Uri.

From the boat stop at Brunnen, which you can reach by car or train, it is an hour by boat to Fluelen with a stop at **Rutli**, which is the sacred meadow where the men of the first three forest cantons swore

Kapellbrücke, Luzern

Heartland Switzerland 101

to stand by each other and found the Swiss confederation. There's a stop, too, at **Tellsplatte** where there is a chapel on a superb, isolated site recalling the best legend of William Tell. Tell was being taken by boat as prisoner of the dreaded Gessler, after the incident of the apple, when a storm sprang up. Gessler had to ask Tell to take charge of the boat. Our William steered the boat to a rocky ledge, like that on which the chapel stands, leapt ashore, and kicked the boat back into the stormy lake.

These spots can be seen from the **Axenstrasse**, the road on the opposite side of the lake which is a corniche as far as Sisikon, and then enters a tunnel leaving the old gallery road to be followed on foot. The Axenstrasse gets frantically busy in summer.

A ring of quiet resorts surrounds the lake; all of them give immediate access to gentle and severe mountain walks, and boating on the lake. The larger include: **Alpnach**; **Brunnen** – along the warmest stretch of the lake, a few highish-rise hotels, plus older chalets by the lake, best starting point for the historic view of Lake Uri; **Gersau** – minute, pretty church, steep hills down to lake, cable car; **Hergiswil** – above here you can take a cable car to the Seeblick lookout (724m/2707ft); **Immensee** (on Lake Zug, not on the Vierwaldstättersee) pretty, peaceful, activity centre in a more pastoral setting; **Kussnacht** – on gentle slopes, town-chalet style main street, cable car to Seebodenalp for skiing; scene of the Hohle-Gasse, the gorge where Tell got Gessler in the end; **Seelisberg** – 400m/1312ft above the lake, surrounded by meadows and mountain woodland hiking centre, small boating lake; **Sisikon** – small and quiet with mainly modern houses, restful but backed by the busy Axenstrasse; **Stansstad** – yachting centre squeezed in a strip of meadow between the lake, the autobahn and the wooded hill; sandy beach; cable railway from Stans to the Stanserhorn, a green viewing point at 1900m/6234ft; **Vitznau** – backed by steep mountains, sandy beach, newish hotels, campsite – the poor man's Weggis; **Weggis** – the prettiest and clearly the most expensive of the lakeside resorts; full range of socializing activities, splendidly warm climate.

Meiringen K11

Bern (pop. 4000) A comfortably compact little town, without the scattering of chalets on the outskirts that usually means a tourist centre, but in fact a popular excursion base. The Reichenbach Falls, 250m/820ft up from the funicular station just outside the town, are the 'dreadful cauldron' into which Sherlock Holmes and Professor Moriarty were hurled to their deaths by their despairing creator. A short walk from the funicular station is the entrance to the Aare Gorge, a 1400m/4593ft long cleft in the rocks, nearly 200m/650ft deep and in places only 3m/10ft wide – you walk along the gorge just above the rushing river on a wooden walkway or through tunnels to the second entrance with car park outside **Innertkirchen**.

Chairlift to **Planplatten** for winter skiing or for the little Engstlen lake; bus to **Rosenlaui** for good climbing and the tongue of the glacier; starting point for the round trip by bus over the three passes (Grimsel, Furka for the Rhône glacier, and Susten).

Muotathal G16

Schwyz A simple village, cut off, very alpine. Popular for canoeing and for caving; baby ski slopes.

Sarnen H12

Obwalden (pop. 7300) Old-fashioned town on Lake Sarnen, family centre for hiking and camping, in winter for cross-country skiing. All around the lake there are all the ingredients of alpine scenery – water, alpine meadows, orchards and forests, rocks, all overlooked by the high mountains, but less dramatic than in the better-known places of the same sort. Other towns in the district are similar – **Sachseln** with its old abbey, by the lake. **Flueli**, now joined with Ranft, is holy ground for the Swiss because this was the home of Brother Klaus, a hermit whose wisdom saved the Swiss from civil war in 1481 and provided a balance of pressures by which the Swiss still live. **Lungern** with its little beach on Lake Lungern; **Kerns** and **Melchtal** with its deep gorge. Only **Melchsee-Frutt** (1920m/6227ft) by its bleak lake is created purely as a resort for skiing.

Schwyz G15

Schwyz (pop. 12,000) A quiet, old, by-passed little town, but in a sense the capital of Switzerland. The town and canton gave their name to the whole country, the flag of Schwyz (white cross on a red ground) has become the national flag, and the sacred documents recording the growth of the country by a series of pacts between fiercely independent little states are housed here in the Federal Archives. The elaborately built houses in the old town are the work of old soldiers of Schwyz in the days when Swiss mercenaries were richly paid by foreign princes

– they returned to their old home laden with rewards of courage. Shut in by the mountains and overlooked by Mt Mythen, nothing much has happened there since. Its winter ski resort is **Stoos** (no cars), from which a cable car goes to the viewing point of the **Fronalpstock**; the galleries of the **Holloch** grotto, biggest in Europe, about 15km/9mi away.

Sempach E11

Luzern (pop. 1500) Its main square looking unchanged since the Middle Ages, Sempach was the scene of one of the great battles that made Switzerland, when in 1386 the confederates defeated the Habsburgs and killed Duke Leopold III of Austria. **Sursee** is a similar town on Lake Sempach, with its town hall, Renaissance houses and red-tunicked grenadiers. The district is in the pre-alps, surrounded by low meadows and gentle wooded hills, always within sight of the majestic Alps.

Simmental L5

Bern The lower valley of the Simmen, from Spiez up to Zweisimmen, is quietly pastoral, home to the brown Simmental cattle, but the upper valley and the valleys branching off it house a string of ski resorts which, in summer, are all within walking distance of each other. **Gstaad** is the most fashionable, and like Crans and St Moritz attracts the minor royalty and stars and beautiful people, with a heliport to smooth their way. It also has the most beautiful position, at the junction of four valleys. The hotels – few of them less than three star – are modern town-style, whereas **Saanen**, 3km/2mi away, consists of wooden chalets. **Saanenmöser** and **Schönried**, down the valley from Gstaad, are more scattered in the heart of ski-lift country. In the upper Simmental, **St Stephen** lies across broad meadows and includes a number of different little villages, while **Lenk**, further up in a narrower part of the valley, is nearly as big as Gstaad and is also a spa, with sulphur springs. Even **Zweisimmen**, which is at the start of this whole pleasure ground known as the 'Green Highlands', has its aerial gondolas to take skiers up to the Rinderberg so they can slide back to their chalets.

The Green Highlands are a summer playground for hiking in alpine pastures and wooded mountains, or climbing mountains, with riding, heated swimming pools, fishing, golf, fitness tracks, tennis, camping, and some unpredictable etceteras like the Menuhin music festival (Gstaad, August); in winter it is devoted to skiing (the ski lifts have a total capacity of 30,000 people an hour, if you can imagine that lot all coming down at once) with the usual accompaniments like ice-skating, curling championships at Gstaad, fondue parties, and hospital beds.

Thun J6

Bern (pop. 33,000) Where the river Aare flows out of Lake Thun, a long, narrow island is covered by the narrow streets of the old town of Thun, mostly simple shops with some smart boutiques. Bridges – there seem to be dozens of them – connect the island with the modern town and station on the south side of the river, and with an older quarter on the north side where the castle contains the historical museum and overlooks the medieval town hall square, still used as a market place. In the main street cars occupy the road but pedestrians can walk along the roofs of the shops, on a terrace-footpath. Pleasure steamers which ply Lake Thun come right into the town, along the Aare canal.

The lake shore is fairly solidly built up for a few kilometres either side of Thun, with villas, landing stages for the steamers and sailing boats, little villages; then it gives way to cultivation. On the north side, the ground rises fairly sharply with vines here and there and there are a number of resort-villages like **Hilterfingen** (strong on water sports), **Merlingen** (more water sports, walk to the stalactite caves) and **Beatenbucht** from which a funicular leads to the inexpensive ski-resort of **Beatenberg** which is sunny and peaceful. These places are noticeably less pricey than many. The south shore of the lake is flatter, with the high Alps in the background; chief place is **Spiez** on a picturesque bay with a little marina and medieval castle.

Zug E14

Zug (pop. 24,000) Lake Zug is a peaceful, undramatic lake, just big enough to run to a lake steamer. At its southern end, like Luzern's lake, it is hemmed in by mountains but for the most part lies among gardens and orchards. The town of Zug is spread along its north shore, backed by wooded hills, with an attractive lakeside promenade. It's a semi-industrial town with a modern shopping area that's clean even by Swiss standards, but small enough for the old part of the town, with oriels and overhanging roofs, to be still prominent. The remains of the fortifications may still be seen, the painted houses round the fish market, the gaily-tiled time-tower and the old houses around the Rolinplatz. Cable railway to the **Zugerberg** (990m/3248ft) for an all-round view and winter skiing.

NORTHERN SWITZERLAND

Between the Rhine and the Jura mountains to the north and the high Alps to the south is the region the Swiss call the Mittelland, and mis-translate as the plateau. It's not really a plateau, and certainly not flat, but it's low-lying and subdued in comparison with the rest of Switzerland. In this northern region, with less than a quarter of the land area, live over half the people of Switzerland. They are concentrated in the big towns, of course – Basel, centre of the chemical industry, and Zürich, with Winterthur, St Gallen, Baden, Olten, for heavy manufacturing industry.

Where the land is flat the towns sprawl out into the country, with warehouses visible through the trees and spruce trim factories in the green fields – no besmirching chimneys, for it's all electrically powered. But most of the countryside is gentle or steep hills, wooded, still a rural escape for city dwellers.

In the east, cantons St Gallen and Appenzell, are the pre-alps of farmland interspersed with industry, rising to the fiercer Säntis and Churfisten ranges, high enough for some winter sports; the colourful Appenzell hills and the peaceful valley of the Toggenburg are the real tourist centres here. South of the Bodensee (Lake Constance) is a fairly level landscape of apples and quiet farms, which is repeated north of Lake Zürich, with more hilly country in between. This is repeated in the little enclaves of canton Schaffhausen, and just outside the busy town of Schaffhausen itself are the Rhine falls, a favourite destination for honeymooners.

Near Basel begin the foothills of the Jura, deeply cleft ridges with a string of major manufacturing towns where the hills meet the more open country of the Berner Mittelland. This is typical country of the Swiss 'plateau' – rolling farmland with patches of wood and patches of development, rich towns with a medieval flavour, especially the capital Bern, giving way to more hilly country, where the brown cattle browse, in the foothills of the Alps. Always within sight are the high Alps of the Berner Oberland.

The region is covered by a full network of fast modern roads and motorways, but usually the old road still exists within reach of the new, and if you've the time to follow this there's plenty to be seen of a Switzerland you might think has vanished

– old barns and farmhouses mostly in the Bern style, with curving eaves under the broad overhang of the roof, isolated churches, ruins of castles on the hilltops (for this rich area is what the feudal nobles sought to control, not the empty mountains). Most of the modern housing and other building in the plain is neat, clean and rather formal.

Aarau   I9
Aargau (pop. 18,000) The Old Town, alongside the river Aare, typifies a Swiss town of the plains – overhanging eaves with gaily painted undersides, chairs and tables out on the traffic-free cobbles. The surrounding modern town is busy with textile machinery, shoes and optical instruments, and preserves the old tradition of bell-founding. Aarau is the capital of canton Aargau which is being built on and industrialized wherever the ground is flat but elsewhere retains the undisturbed look of the 17th century – houses shuttered, with blank gables square to the roadway. The Upper Aargau, south of Langenthal, and the old town of Wangen with its wooden-covered bridge, are the most attractive parts of the canton.

Altstätten   J20
St Gallen (pop. 10,000) Between the Rhine and the abrupt climb up to Appenzell and the Alps is a strip of intensely fertile land whose main road is an almost unbroken string of modern little towns, scruffy by Swiss standards. Then suddenly there is Altstätten, retiring, medieval, its market for unrich Swiss.

Appenzell   J20
Appenzell Innerrhoden (pop. 5200) '*Mein Vater ist ein Appenzeller, Er isst den Käs mitsamt den Teller.*' So runs the old rhyme, reminding the good people of Appenzell of their past poverty. 'My

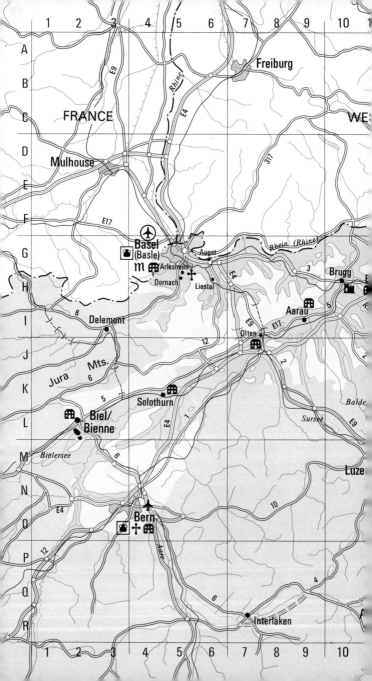

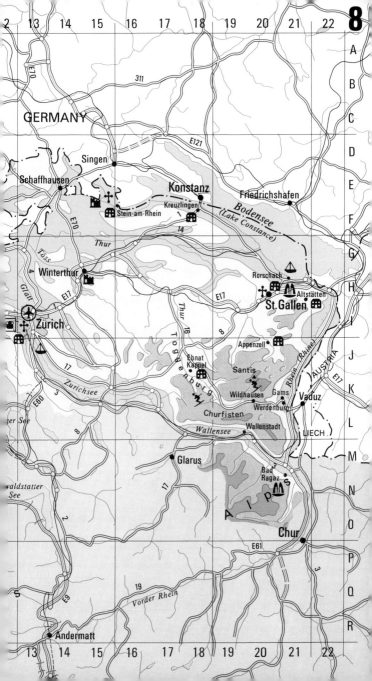

father is an Appenzeller, He eats the cheese and then the platter.' Even today, the Inner District of Appenzell is the poorest canton of Switzerland and this is slightly reflected in the prices. The surrounding countryside is a land of quiet green, steep hills with neat, out-of-the-way farmhouses, but its chief delight is in the towns, Appenzell especially. Pastel-painted houses in broad quiet streets, positively aglow at sunset; the local costume is more garish, but comes out only for photographers; the real local costume is just the fashion of fifty years ago, with an umbrella to represent the ancient right to a sword, and, of course, to keep the rain off. But the umbrellas are often closed.

The main square, used most of the time as a car park or market place, is the scene once a year (last Sunday in April) of the *Landsgemeinde* – the people's assembly when all the citizens of the Catholic half-canton of Inner Rhodes (men only, of course) gather to vote on the cantonal laws and taxes. The umbrellas are replaced by swords or bayonets, the badge of the right to vote. For the Protestant half-canton of Outer Rhodes the *Landsgemeinde* is held in odd-numbered years at Trogen, in even-numbered years at **Hundwil**, flowing out of the square and up the hill. Other towns of the Appenzell district are similar to the main town though less colourful; **Trogen** on its hill is scattered round a minute nucleus and nearby is the first Pestalozzi children's village; **Herisau** with 15,000 inhabitants is a substantial town with some palatial houses and entertainments, but strives hard to retain the feeling of a village.

The tradition of embroidery, from which arose the textile industry of St Gallen, survives as individual handwork in the Appenzell district. **Lake Seealp** overlooked by the cliff-like start of Mt Säntis is a good outing from Appenzell.

Baden H11

Aargau (pop. 13,600) A warm-spring spa, with the parks, gardens, riverside terraces, pools and casino to complement the cure-hotels. Shielded from the spa quarter by a bend in the river is the industrial town, busy with electrical engineering, while the more picturesque Old Town is up-river.

Basel G5

Basel-Stadt (pop. 200,000) Basel has the highest income per head of any city in Switzerland, probably of any city in the world (outside the oil states). The wealth doesn't show – it looks just prosperous, comfortable, cultured, not so sober as Zurich but just as sensible with money. It's a major port on the Rhine, an insurance and banking centre, and here are the headquarters and laboratories of the major Swiss pharmaceutical companies – Roche, Ciba-Geigy and Sandoz, their oldest machine rooms temples of cleanliness like brand new hospitals. Squeezed up tight against the frontiers with France and Germany, the factories sometimes have to spill over to a neighbouring country, with two lots of police and factory inspectors within the same wire fence. St Louis, the suburb which lies across the border, looks totally French.

What most visitors see first is the Old Town, gathered on the hill round the Gothic Münster (not a cathedral – it was not the seat of a bishop) where Erasmus is buried. The surrounding square, with its Alsatian-timbered houses, is one of the most agreeably proportioned of all, and its Cafe zum Izark an introduction to Basel's young life. From a terrace off the square you look across the Rhine down to the other old town of Kleinbasel, traditional rival of the big city. This rivalry can be seen most at Carnival time, in February, and if you want to hear an argument, enquire whether the alleged obscenities of carnival still take place.

Chief shopping street, just below the Old Town, is the Freie Strasse; at one end is the ancient market place, still used daily for fresh produce, and overlooked by the 16th-century town hall and guildhall. The fish market is a little further on; near the other end of Freie Strasse is the centre of Basel life, the Barfüsserplatz. The Historical Museum is on this Platz, and the New Theatre with its amusing water fountains and the Hall of Art are a few steps away. But the most visit-worthy of Basel's museums is the Kunstmuseum (Fine Arts) with one of the finest European collections, constantly changing special exhibitions, and very strong in impressionists. Holbein worked in Basel and is remembered in the Holbeinstrasse which runs from near the zoo to Holbeinplatz.

Outside Basel, visits are according to taste. There's the Roman ruins at **Augusta Rurica**, with an 8000 seat theatre where plays are given in summer and a faithfully reconstructed Roman house. Alternatively there's the rich suburbs, with their discreet, unshowy separateness, tidy houses in unloved, railed-in gardens, with height the only clue to price. Two of these suburbs are **Arlesheim**, with a delightful 17th-century church, and nearby **Dornach**, where the anthroposophist followers of Rudolf Steiner maintain their spiritual university round the curiously shaped Goetheanum.

Bern 04

Bern (pop. 150,000) Bern takes its name from the bear which was the first animal caught by Duke Berchtold after founding the city, and the bear is still seen on the flag not only of Bern but of many places that were once part of its little empire. The only live bears left now are those in the bear pits (No. 12 tram from the station to the Nydegg bridge) waving so friendlily to visitors for food, but very aggressive to birds that fly in to join the feast.

The days of empire are past, and Bern does not wear the prosperous industrial uniform of other major Swiss towns. But it remains the federal capital, with the patrician feel of a natural capital (but since Switzerland does not have a central government, there is no horde of civil servants to make it a bureaucrats' city). The old town is enclosed by a U-shaped bend of the river Aare. Its main street, which runs the length of the middle of the old town under four different names, is lined with lived-over shops whose fronts are sandstone arcades which give the city much of its grace and charm. The back streets off the main road are similarly arcaded, with extraordinarily expensive antique shops next to old-fashioned workshops and repair sheds. Steps lead down to the river level, where tucked away, cheaper eating contrasts with the more celebrated Bernese style of eating place, in converted cellars of some of the old buildings like the granary on Kornhausplatz. It's a low-key city – solid, reliable, but if you accept it as that, extremely agreeable. That's why the Swiss let it remain the capital.

The legations and embassies are concentrated around the **Helvetiaplatz**, across the river from the old town by the Kirchenfeld bridge from which there is a splendid view down to the weir and across to the spire of the **cathedral**, Switzerland's highest at 100m/328ft. In this sedately wooded region, too, are some of Bern's museums – the postal, the Swiss national library, rifle, historical and natural history museum, and one which is a must for the keen climber – the **Alpine Museum**, showing the development of equipment and refuges and with a superb collection of relief maps, which can teach the serious climber a lot. Here too is the **Swiss Patent Office**, where Einstein worked as an examiner while developing the Special Theory of Relativity. The **Langgass quartier** is Bern's little Latin quarter, where a lively mixture of nationalities congregates.

The **Fine Art Museum**, very good on Swiss primitives and Paul Klee, is on the opposite side of the old town over the

Clocktower, Bern

lower loop of the Aare. Other things to see are: the **Clock Tower** in the main street, with a much-photographed mechanical puppet show three or four minutes before it strikes the hour; the **Parliament** building with conducted tours – very instructive in another side of the balanced working of Swiss democracy; the market on Tuesday and Saturday mornings in the **Bärenplatz** behind parliament; the late-medieval **Town Hall** with a delightful council chamber in a lovely little square; the **Rose Garden** up a steep climb beyond the bear pits – perhaps this should be the first thing to see, for its lovely view of the whole city.

Biel/Bienne L2

Bern (pop. 58,000) An industrial and commercial town, mainly square and level but with a charming 'old town' on a low hill above the new market place. The old town was extensively done up in 1936 and it is difficult to tell what is genuine and what reconstruction, but it doesn't matter – the effect of the short High Street and the two little squares round the 17th-century Town Hall and 15th-century church is faithful and charming.

As in Fribourg, French and German are the languages of Biel, but unlike Fribourg there is no language frontier, the people are bilingual, speaking both with equal ease (or, as purists observe, with equal difficulty). Being in canton Bern, its German name Biel is usually put first, but on trains and buses the German name and the French name Bienne are often run together to become one.

Biel itself lies in the Swiss Mittelland, at the very foot of the Jura mountains. A short 6km/3½mi into the mountains brings you to the **Taubenloch gorge** (a deep defile with good footpath cut in the limestone); a longer expedition round the flat side of Lake Biel, marshy in places, brings

108 Northern Switzerland

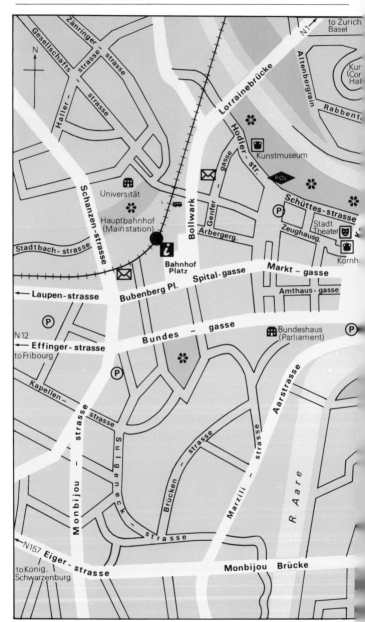

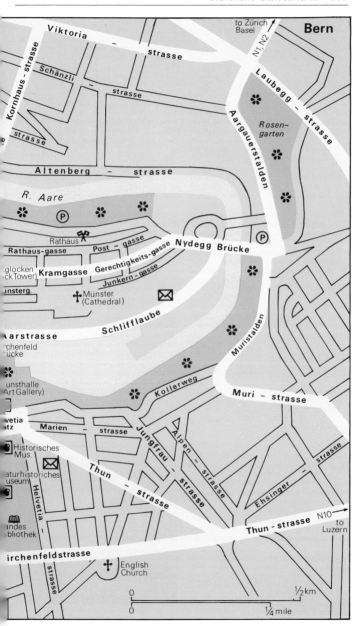

you to the causeway leading on to **St Peter's Isle**, a peninsula in Lake Biel to which Rousseau and many after him have fled for complete peace (also by boat from Biel or from La Neuveville, p. 123).

Brugg H10

Aargau The sedate old town of Brugg is entered by a pedestrians-only bridge across the Aare from the main road; very quiet up its steep hill in contrast to the surrounding modern town. About 6km south of Brugg is the Habichtsburg, castle of the hawk, the original castle of the Habsburg family whose rise and fall is the story of Austria and Switzerland.

Kreuzlingen F20

Thurgau (pop. 17,000) This is a suburb of Konstanz in Germany with which it forms one town though divided by a frontier with customs posts (passport needed). There is a picturesque small old quarter in Kreuzlingen, but the interest is in the very Swiss-looking larger town, where still stands the Konzilgebäude where the Council of Constance held its sittings in the late 15th century, deposing a couple of popes and burning John Huss for preaching against the pope. Konstanz stands at the end of the Bodensee (Lake Constance in English) where the Rhine flows out of it, and is a major port for traffic on the lake. The lake front, with gardens and a real casino, is very agreeable.

Olten I7

Solothurn (pop. 24,000) A great railway-building centre, plus cement works and soap factories along the river. Olten retains a medieval core on the left bank, by the covered wooden bridge, and in its modern residential parts is typical of the Swiss art of making life comfortable if unspectacular.

Bad Ragaz N20

St Gallen (pop. 4000) A spa town right in the mountains, now also quite prominent as a base for skiing in the Alps up to Pizol (2895m/9500ft). Good outings from here are to the **Tamina Gorge** (at one time people were lowered into the gorge by rope to 'take the waters') and the **St Lizisteig** defile, and to the village of **Maienfeld** (canton Graubünden) where Johann Spyri wrote and set his famous children's novel *Heidi*.

Bad Ragaz is a good starting point for the Walensee, a fjord-like lake in wild country with a number of growing ski resorts like **Amden**, **Walenstadtberg** and **Flumserberg**; 3km/2mi aerial cableway to the Alp Tannenboden (1400m/4593ft).

Rorschach H21

St Gallen (pop. 10,800) The largest port on the Bodensee, good boating centre and base for outings on the lake; mainly a business quarter, simple local museum. The countryside along the south side of the lake is typical of the Swiss Mittelland – small hills crowned by castles, apple orchards and vineyards for the good red wine, small pine forests separating the small meadows. From Rorschach there is an immediate outing into the **Rorschach Berg**, or longer one by cogwheel train to the small spa of **Heiden**.

St Gallen H20

St Gallen (pop. 76,000) An ancient abbey town, now prominent as the major textile centre of Switzerland – noted for designs and for the smoothness of the machinery. Much of the town is now huge admin blocks and modern buildings like the New Theatre and the University, but there is a large area of old town, gathered round the cathedral; visit especially the library of the former abbey. The **Deer Park** has ibex, chamois, and wild boar roaming loose – much easier to see them than in the National Park, even if less fun looking for them. October is the time of the OLMA, a fair for the agriculture and milk industries, when the streets are a sea of banners.

Schaffhausen E14

Schaffhausen (pop. 34,000) A big industrial town, busy with aluminium smelted by hydroelectric power from the Rhine – the town took its name from the 'ship houses', where cargo was stored when ships had to be unloaded for their goods to be carried past the falls and rapids. There is an attractive old town, one of the best preserved of Switzerland's many medieval towns, with jetties by the river overshadowed by Munot castle (circular keep as theorized by Dürer), curving

Rhine Falls, Schaffhausen

Munot Fort, Schaffhausen

alleys with clocktowers and statues, and the brightly muralled Knight's House. Romanesque Minster, with peaceful three-storeyed cloister. But more than for its old town, Schaffhausen is visited as a starting point for the Rhine Falls, just below Neuhausen 5km/3mi downstream. The river is 150m/137yd wide at this point, and falls 23m/75ft over the rocks in the course of about 50m/54yd. In July, when the melt-water from the Alps reaches the falls, the flow is more than 1000cu.m/sec (Niagara: 6003cu.m/sec). Hotels overlook the falls, and boat trips across the river.

The landscape of canton Schaffhausen, two little enclaves north of the Rhine, is blue with grapes or yellow with barley and mustard, but becoming more wooded on the hilly ridges, as agriculture has a difficult time in this part. **Schleitheim** is one of the most characteristic, half-timbered, bendy villages.

Solothurn K4

Solothurn (pop. 18,000) One of the oldest towns north of the Alps, though now just a comfortably modern city. The old town, a paved pedestrian precinct and shopping centre, lies between the Basel gate and the grim Biel gate. A favourite outing for the locals is up to the **Weissenstein** – a twisting gravel-surfaced road to a peak above the fir trees, with views across to the Jura. There's a short-cut by chair lift from **Oberdorf** (4.5km/3mi).

Stein am Ruein   **F16**
Schaffhausen (pop. 3000) The gaily painted fronts of the houses in the old town, timber-framed and a jumble of oriel windows, and the relative absence of a bustling modern town all around, make Stein one of the most picturesque medieval towns in Switzerland. This stretch of the Rhine is rich in old monasteries and castles.

Toggenburg J17

St Gallen This is the upper valley of the river Thur, with **Mt Säntis** in the north (2504m/8215ft) and the jagged **Churfirsten** range to the south. It has remained a pastoral backwater, rather like the Appenzell district but more mountainous. The valley is almost a parkland of picturebook boarded houses and cattle, in the mountains goats and plenty of skiers. Best equipped ski centre is the scattered triple village around **Wildhausen**. Further down the valley **Ebnat-Kappel**, with its detached square stone houses with steps leading down to the road, is quite individual. The small Calfeisen valley is notable for wildlife, while the time to see the cattle is at the end of summer when they are brought down to winter quarters in the ceremonial Alpabfahrt.

The winding descent from the Toggenburg through woodland to **Gams** is very attractive. Nearest large town is **Buchs** with its shopping centre and tower blocks sprawling over the green slopes like the brashest ski-resort; **Werdenberg** just outside Buchs is a village recalling the more arduous past, still surveyed by its grey castle.

Winterthur H15

Zürich (pop. 87,000) Most of the mountain railway engines you see in Switzerland seem to have been built in Winterthur, which is a predominantly engineering town in a rural setting. The people like to live comfortably and enjoy themselves, and you can join in. Good music, outing to **Kyburg Castle** for collection of musical instruments.

Zürich I13

Zürich (pop. 380,000) This is the largest town and financial capital of Switzerland; the gnomes of Zürich live along the Bahnhofstrasse, in the giant blocks of the great banks, rising to no great height so you don't realize quite how vast the banks are until you see that one building is connected to the next and the next, interlocking complexity like Swiss banking itself. But the secret of Swiss banking is not so much its secrecy as its probity, a vice that seems to affect all the people of Zürich.

Zürich rather than Geneva is now the Protestant Rome – a sober, puritan city, but very kind, friendly and able to be hospitable unless your idea of fun is to stop out after 2 a.m. In its retiring way, it's the liveliest city in Switzerland, and also the only Swiss city that has shown signs of the revolt of European youth against the comfort of its elders.

Industry, saved from grime by reliance on electricity, is concentrated in the northern suburbs; the city centre is shops, commerce and formal entertainments. There is more sense of fun around the lake; the town stands at the end of Lake Zürich (Zürichsee), where the river Limmat runs out of it, and the two quays are used for relaxation. While the left bank of the Limmat houses the business section of Zürich, the right bank is for culture. Here are the main theatre, the Kunsthaus, (fine arts museum, constantly changing displays) and the cathedral (Grossmunster) and further away from the river the staid university quarter. Zürichberg Strasse, which leads away from the university to the zoological gardens, is where the Quartier Latin may be found.

The Swiss National Museum is the one most worth visiting; in a park right by the main station, its individual items look museum-like and dead, but the overall effect, especially in the collection of costume and prints, is to show how Switzerland evolved and something of what the Swiss feel.

There are boat trips on Lake Zürich, but it is more interesting to tour round it by car. For about half its length on both sides of the lake there is an untidy sprawl of tidy little suburbs, quite expensive, very select. Suddenly the urbanism stops and you are in a trim orchard, simple lowland farming with 20th-century comfort in 18th-century tranquillity. **Rapperswil** near the east end of Lake Zürich has a medieval core in its upper town, overlooked by the castle in a deer park.

Zürich

WESTERN SWITZERLAND

Predominantly French-speaking, this region of the country was the popular picture of Switzerland in the days before Alps and snow became tourist attractions. No yodelling, no William Tell, no gnomes, but totally Swiss in its precision, cleanliness, love of freedom and pursuit of order.

Canton Vaud was a subject territory of Bern until Napoleon made it a separate canton, and like Bern it is Protestant; the farmhouses and the street-scenes of the old towns look Bernese too. Canton Fribourg, a member of the Swiss confederation since 1481, has remained solidly Catholic as though in contrast to Bern, and predominantly rural; one third of the canton is German-speaking. Neuchâtel became attached to the confederacy at about the same time as Fribourg, but a succession of royal rulers to protect its Protestant faith left it with the king of Prussia as its prince, even after becoming a canton in 1815; you may still see the royal arms on public buildings in Neuchâtel. The tiny canton of Genève, which is little more than the land around Genève city, was an independent, fiercely Protestant, republic until it joined the confederation in 1815. The canton of Jura was part of Bern until 1978 when this Catholic, French-speaking area seceded from the Protestant, German Old Canton after years of conspiracy and petty terrorism – you may still see the slogans on public walls in Jura.

For the tourist, there are four distinct parts to this region. In the south, Lac Léman (which you can call Lake Geneva if you like, but not in Lausanne) stretches

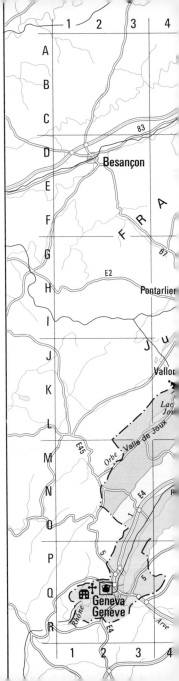

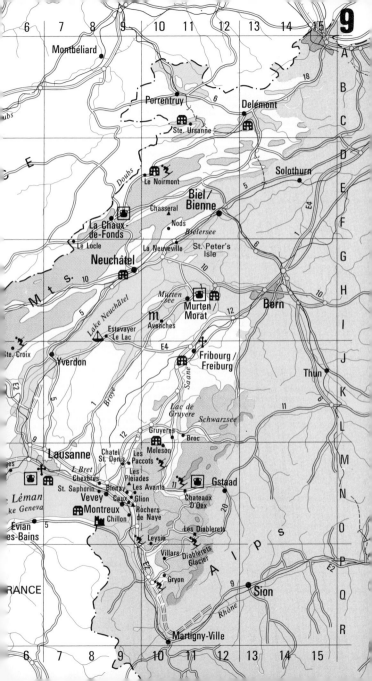

for eighty miles from cosmopolitan Geneva past industrious Lausanne and laze-away Montreux to the little castle of Chillon, and all along its north shore the hillsides are a dense patchwork of wheatfields and vines, their terraces looking across the lake to the highest of the Alps on the French side around Mont Blanc. This part is sometimes called the *Vaud Riviera* for its mild climate and sunny atmosphere. Steamers ply the lake in a regular service calling at the cities, small lakeside towns like Rolle and Morges, and dusty little villages. Inland, a twisting corniche road winds between Lausanne and Montreux, from which you reach vineyard villages like Chexbres in the hills or St Saphorin on the lake.

The Jura mountains form the barrier between Switzerland and France in the west of the region (canton Jura is only a small part of these mountains). Long valleys, intensely cultivated or pastured, some of them filled with twisting streams and gorges, lie between ridges of hill which rise steeply through dense woods to clear plateaux where rural life goes on with hardly ever a foreign visitor. The few towns are devoted still to watch and clock making; the villages are poor by Swiss standards but beginning to prosper now from a growing interest in horses and winter sports – Swiss come here for the cheaper skiing, especially cross-country skiing. The highest point, Le Chasseral, is only 1600m/5249ft. The western boundary of the Swiss Jura is marked by the winding river Doubs. The old spelling of the name – Doux meaning sweet – was more appropriate, for it is a very pretty river, in a deep gorge.

The eastern boundary of the Swiss Jura is a line of lakes – Biel, Murten, grand Lake Neuchâtel and little Lake Joux – and between the lakes and the Alps is the rolling green countryside of the Swiss Mittelland. The Swiss normally translate this as 'plateau', but it's hardly flat – in places as hilly as the highlands of Scotland, and everywhere in sight of the foothills of the Alps. Medieval towns, Fribourg especially, are the chief attraction for visitors. The brightly coloured statue of a well-endowed young lady, blindfold, that stands in many towns, represents Justice. The most visited village is Gruyeres, where the cheese comes from. Just a small part of this region, east of Montreux, is alpine or pre-alps. Leysin, Villars and Chateau d'Oex are three places that were villages not so long ago, and are now substantial ski resorts and summer hideouts, not yet with the high-rise hotels of some other places in the mountains.

Château d'Oex N11
Vaud Chief resort of the Pays d'Enhaut, regularly laid out in a wide valley, with ski lifts up to 2350m/7710ft. Friendly rather than swish. Museum of the rich past; outing to an attractive cleft called the Pissot Gorge.

La Chaux de Fonds F9
Neuchâtel (pop. 45,000) Largest watch-making town of Switzerland, built to a very straight-lined grid in the early 1800s and now surrounded by mid-1900s skyscrapers. Excellent museum of the timepiece industry. Nearby Le Locle is another, smaller, watch-making town where the first Swiss watch was made; near the cliffs and waterfall of the river Doubs.

Chexbres N8
Vaud (pop. 1700) At the highest point of the corniche between Lausanne and Montreux, centre of a rich wine-growing area overlooking Lac Léman. Fishing on the small Lake Bret, winetasting in Caveau des Vignerons.

Les Diablerets P11
Vaud A visit to the glacier at 3000m/9842ft is the attraction of this village if you are not a skier (reached in three stages by gondola and then cable car fron Col du Pillon). The village of scattered chalets and solid hotels lies in a broad valley looking up to peaks of perpetual snow.

Delemont C13
Jura (pop. 12,500) A watch-making centre, chief town of canton Jura, retains its old town above the river – mainly classical 18th-century buildings. The road between Delemont and Porrentruy across the low Rangiers pass, is an established scenic run, but the winding road to Laufen and Basel along the river Birse is more intimate and attractive.

Estaveyer le Lac F9
Fribourg (pop. 2500) An old walled town on the south shore of Lake Neuchâtel, very good for water sports. **Payern** (Vaud) 10km/6mi east is another fine old town, with a simple (restored) Romanesque abbey church of the 11th century.

Fribourg/Freiburg I11
Fribourg (pop. 43,000) The river Sarine, the traditional boundary between Alemmanic (German) and Burgundian (French) culture, loops in a deep gorge round the city of Fribourg, whose old town lies down by the river; little squares with tables outside, narrow cobbled or stepped streets, high dark medieval

Fribourg

houses and remnants of the old ramparts (Tour Rouge and Tour des Chats). Fribourg has a long and proud history, from its princely foundation in 1157 at a ford in the river, through life as an independent republic until it joined the confederation, and persistence as a Catholic stronghold when its neighbours turned Protestant. This is seen today in the many churches and in the maintenance of the blue-uniformed Fribourg Grenadiers, trotted out on grand occasions as though off to resist Napoleon. The chief churches to see are: the cathedral of **St Nicholas**, a sandstone building dating from about 1300, and the **Church of the Woodcutters**, a Franciscan friary dating from the end of the 13th century with lovely painted altarpiece and carved wooden triptych. The **Town Hall**, dating from the 16th century, stands in the upper part of the town in front of a square still used as a market for local produce on Wednesdays and Saturdays.

The best views of the old town, looking down on the red roofs of the dignified, very lived-in houses, are probably from the bridges – the **Zähringen Viaduct**, **St John's Bridge**, and especially from the arches of the **Gottéron Bridge**.

Genève Q2

Genève (pop. 152,000) Genève (in English, Geneva, in German, Genf) you may be told, is not really Swiss – it's French, or international, or uniquely Genevois, but not typical of Switzerland. Don't believe it. True it's cosmopolitan – one person in three is a foreigner, but most of them stay in their enclaves and anyway, one person in six in Switzerland is a foreigner. It's distinctly French of course – at the back of the smart modern buildings there are still old poorer blocks that might have been brought here from the back of Paris' Gare du Nord. The food is very French, too, though there's every other variety of cooking for the international set. And of course the town still looks (in its old part) and acts like an independent republic – isn't that the essence of a confederate canton? But in its sharp cleanliness, the quiet of the busy streets, its order, its enjoyment of fun – restrained but uninhibited – Geneva has everything that Switzerland is accused of.

Geneva was called the Protestant Rome when Calvin installed his puritan government there in 1541, but it's Rome no longer – no vice, no scandal, no dolce vita. But it tries, with the only serious attempt at nightlife in Switzerland.

The city stands at the western end of Lac Léman, where it narrows to become again the river Rhône which divides the southern part of Geneva, where the old town and the sights are, from the northern, which houses the station, the plush hotels along the lake, and most of the international community. Geneva is famous as the European home of the United Nations and related organizations like the World Health Organization, and UNICEF, housed in the Palais des Nations, which was the headquarters of the old League of Nations and is now open to visitors. Scores of other international groups have their base here – the Red Cross, the I.L.O., World Intellectual Property Organization, International Telecommunications Union, GATT, European Council for Nuclear Research with its giant cyclotron (not open to visitors). In addition to the diplomats who get posted here with no reluctance, there come the refugees, though the days when Lenin could work here, plotting the next step in the Russian revolution, have gone.

Chief connection between the north side and the south is the Pont du Mont Blanc, from which the Rue du Mont Blanc leads up the hill to the station. Upstream from the bridge you look over the lake and its landing stages to Geneva's landmark, the **Jet d'Eau**, a fountain squirting lake water 150m/490ft in the air – it's turned off when the wind is strong, or there would be unwanted soakings. Downstream from the bridge is the river, with an artificial island with a statue to Geneva's most distinguished native, Rousseau. It's a few steps from the bridge to the **Place du Molard**, a centre of young life with outdoor plays and minstrels, hemmed in by two of the smartest shopping streets, the **Rue du Rhône** and the **Rue du Marché** (the bigger department stores are on the other side of the river, off the Rue du Mont Blanc). From here, narrow streets or steps lead up to the old town, to small squares with fountains, cobbled streets, art galleries, and antique shops. Here are the Arsenal with its cannon, the 16th-century Town Hall where the Geneva Convention was signed, and St Peter's Cathedral which started life as a Romanesque Catholic church, became the centre of Protestant preaching in Calvin's time and was re-dressed in the Age of Enlightenment to look like a Greek temple – a most appropriate façade of rationality. Geneva has been proud of its fame as home of the intellect and sciences ever since. On the far side of the old town are the **University**, one blank wall lined with giant statues of the great figures of the Reformation, and its library with some display rooms open to the public, and within walking distance of the university are the museums – **Natural History**, to remind you of the great naturalists who flourished in Geneva in the Enlightenment, and the **Museum of Art and History**, the **Rath Museum** for watches, the **Petit Palais Museum** of French impressionist and later paintings. Other museums in Geneva include the **Ariana** (superb porcelains, HQ of the International Ceramic Academy), **Voltaire**, **Ethnography**, **Musical Instruments**, and out in the **Parc de Mon Repos** the history of science.

Geneva is exceptionally well provided with parks: Mon Repos and adjacent open areas is the largest, like being out in the country; the **Parc des Eaux Vives** with its rose garden at its best in June is a favourite for many people.

The suburb of Carouge, with a street market on Weds and Sats, is almost Bohemian in contrast with the sobriety of Geneva, and no less popular with Genevese for that. There's a casino by the lake on the Quai du Mont Blanc. Real gamblers nip across the frontier to Divonne, or take the boat to Evian-les-Bains.

Most of the year the weather is good in Geneva. November and early winter is the only dreary time – fog and humid cold; livened up by the jollifications of the

Western Switzerland 119

Gruyeres
Escalade on 11 and 12 December, when the citizens recall how their forefathers drove off the Duke of Savoy by pouring cauldrons of hot chocolate over his troops, or some such legend (the facts don't matter so long as you enjoy yourself). An even better time for fun is the Fêtes de Genève, in mid August.

Gryon   P10
Vaud A still-rural alpine village coming into use as a ski centre. **Baroleusaz**, higher up the valley, and **Alpe des Chaux**, still higher, are where the ski lifts begin, Gryon has the accommodation. Nature reserve of Taveyannaz nearby.

Gruyeres   M11
Fribourg (pop. 1200) The very picture of a medieval stronghold village, with its single main street lined with comfortable Renaissance houses winding up to the castle terrace. No cars are allowed within the town from Easter to October but it can

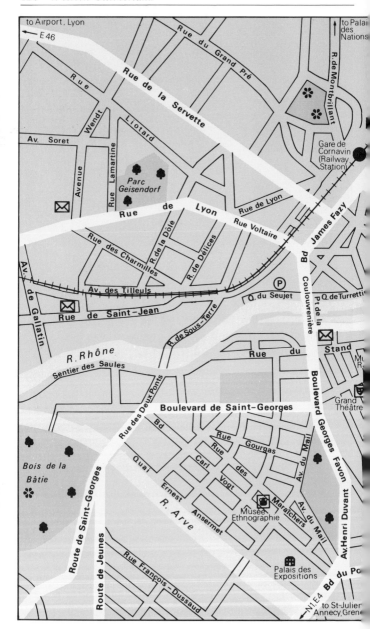

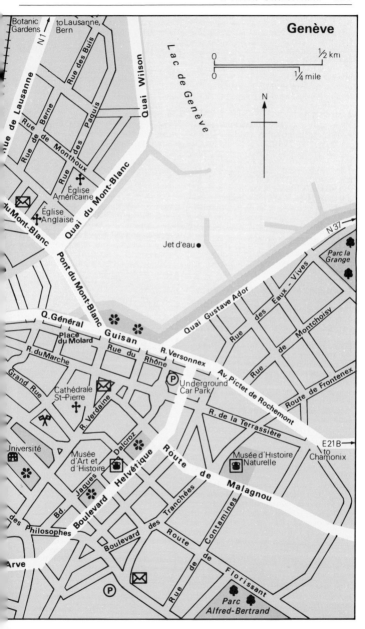

still get pretty crowded in summer. Gruyeres is the centre of a district, La Gruyere, of gentle pastures for quiet cows which produce milk for chocolate made at nearby Broc or the famous Gruyere cheese. Most of the cheese is made today in a factory just outside the village which is open to visitors. The factory is on the road to **Moleson** village, a ski resort.

Lausanne M7

Vaud (pop. 135,000) Lausanne exists, socially and physically, on three levels. Socially, it is a calm, dignified city for the retired rich (out-of-work kings and tycoons used to choose Lausanne for its balmy climate and lakeside views); a bustling business centre and trading post for the industrial towns of canton Vaud; and the lively home of a youthful horde drawn to the university, technical high school, and schools of art – perhaps the most stimulating educational set-up in Switzerland.

The town has grown up on three steep hills, once separated by rivers which have been covered over, today connected by bridges. At the top are the twin hearts of Lausanne, the business hubbub along the Grand Bridge between Place St François and Place Bel Air, and the calmer zone of the old town with the 13th-century Burgundian **Cathedral of Notre Dame** (where the nightwatchman still calls the hours), the Place de la Riponne (where the 20th-century Renaissance-style **Palais de Rumine** houses the university and museums of fine arts, natural history and archaeology) and the old market place, still a flea market on Weds and Sats, called **Place de la Palud**, overlooked by the 17th-century town hall. Luxury and antique shopping in the **Rue de Bourg** and **Rue St François**, cheaper on the west side of the Grand Bridge. Below the upper town, at the level of the railway station, a blend of offices, less high-powered shops, and quiet residential quarters, all quite modern. Below the station, along the lake is **Ouchy**, a 2km/1mi promenade of parks, gardens, yacht moorings and landing stages. The three levels are connected by a steep rack-and-pinion railway (train every 7½ minutes) grandly called the Metro.

Outdoor playlets are performed at Ouchy, also Beaulieu and Municipal Theatres for opera and plays, and one avant-garde theatre. There is a festival of music and ballet in May and June. Favourite nearby outings are to **Le Signal** (1km/½mi north of Notre Dame), **Montrion Park** (below the station) with a splendid viewing point for the Alps across the lake, and **Mon Repos Park** where the Olympic Museum honours the memory of Baron Coubertin who founded the modern Olympic Games here.

Leysin 010

Vaud A modern little town in the Vaud Alps, on a south-facing terrace site overlooking the Rhône valley. Mainly for winter skiing, it has developed a wide range of activities for summer.

Montreux N9

Vaud (pop. 20,000) Montreux developed as an Edwardian resort with a Parisian accent and a mild Mediterranean climate, and there are still plenty of signs of this past in the big, old-fashioned hotels. There has been plenty of modernization, too – up-to-date hotels and shops, casino, swimming pools, municipal gardens, new lakeside quays, plus distractions like the Jazz Festival, Musical September and the Festival de la Rose for television awards. The climate – with figs, magnolias, cypresses and palms by the lake and vines and walnuts high in the hills, making Montreux the centre of the Vaud Riviera.

Montreux **Old Town**, mainly 18th-

Château de Chillon

century houses with the 12th–15th-century church tucked away up the hill.

The **Château de Chillon** must be the most photographed outing from Montreux. It was built to guard the eastern end of Lac Léman by Peter of Savoy in the 13th century, and until the 17th century was a state prison where Bonivard, a prior from Geneva, spent six years chained to a pillar in a dungeon, for supporting the Reformation. Bonivard is the subject of Byron's 'Prisoner of Chillon'.

Other outings from Montreux are into the mountains. **Blonay**, with its 13th-century castle, is reached by a nostalgic steam railway from Chamby, a halt on the fantastically twisting 'main line' from Montreux to Gstaad. **Les Pleiades** beyond Blonay is a winter skiing, summer forest resort with a panoramic view of the lake and many mountain peaks. **Les Avants** is a village with a splendid site overlooking a wooded gorge (Gorge du Chauderon) and near the viewing point of the Sonloup pass. **Glion** is a hillside town overlooking the Gorge du Chauderon and **Caux**, beyond Glion, is a blend of mountain town, ski centre and conference centre for Moral Rearmament. **Rochers de Naye**, ski centre, with all-round view; good for early spring flowers.

Morges   M6

Vaud (pop. 12,700) Wine-growing centre with a little harbour on Lac Léman, 13th-century castle, cosy main street with Wednesday and Saturday market, tulip festival in April and harvest festival in October. The **Alexis Forel** museum has interesting furniture, tapestries, Dürer and Rembrandt engravings, porcelain.

Murten/Morat   M11

Fribourg (pop. 4600) The old town of Murten, on a slight rise overlooking its little lake, seems a miniature Bern with its arcades, overhanging brown roofs and pinnacled gateway. The ramparts (walk round the walls) surrounding the medieval city, and the historical museum, remind you of the warlike past of Murten, scene of the final defeat of the Burgundians in 1476 which ensured the survival of the Swiss confederation. A canal links Murtin's lake and Lake Neuchâtel.

At **Avenches**, 8km/5mi away, are a Roman amphitheatre and other remains of the ancient city of Aventicum, lying neatly at the end of the Rue Centrale, one of the three streets of this tidy medieval village. Roman museum; wooded campsite and sandy beach on Lake Morat.

Neuchâtel   H9

Neuchâtel (pop. 35,000) A dignified, quiet university town, very prosperous as a wine market and research centre for the Swiss watch industry. Sleepy, rather grand buildings down by the lake, where there is a harbour for the steamers of the three lakes and for boating. The Old Town above the lake has 16th- and 17th-century houses, 13th-century university church and graceful Romanesque castle. **Places des Halles** is still a market.

At the eastern end of Lake Neuchâtel is **La Tène**, the site where finds from the Celtic Iron Age (some of them now in the town's archaeological museum) gave its name to the La Tène period in European culture (see Hallstatt, p. 64).

La Neuveville   G11

Bern (pop. 3900) A little medieval town dreaming by Lake Biel, with a cobbled main street and narrow arcaded side streets off it. A starting point for Nods, from where a cable car rises to the Chasseral, highest point of the Swiss Jura with views to the high Alps, the Vosges and the Black Forest.

Le Noirmont   E10

Jura (pop. 1550) One of the prettier villages in the Jura mountains, in a district called the Franches-Montagnes of steep forested hills, green flat pastures for horse-grazing, and long valleys shut in by high ridges. Easy access to the river Doubs and its gorge. Many villages of the Franches-Montagnes are used by the Swiss for winter skiing; **Tramelan** is one of the more developed skiing areas.

Rolle   N4

Vaud (pop. 3600) A scattered little town with a chateau, campsite under the trees by the lakeside, stone-built harbour, surrounded by vineyards. There are a dozen quiet dusty wine-villages in the vicinity, best known is **Bougy-Villars** with its celebrated viewing point, the Signal.

Ste Croix   J6

Vaud (pop. 7000) The home of the Swiss musical box (now nearly defunct) is today a blend of ski resort and manufacturing town busy with radios to replace the musical box. The nearby **Val de Travers** is a valley of attractive, twisting gorges which yielded asphalt and gave its name to the British company, Val de Travers.

St Ursanne   C11

Jura (pop. 1100) A medieval town whose old houses line one bank of the river Doubs, which makes a great loop near here. Entry from the hills across the river is through an evocative 18th-century stone bridge leading to an arched gateway. A pleasant, sleepy place to visit if you're touring this part of the Jura, taking in the **Gorge de Pichoux** (on the main road to Biel) or the **Corniche du Jura**.

Vallorbe   K5

Vaud (pop. 4000) Seen at its best in the autumn, this first stop on Swiss soil of the train from Paris to Milan is an excursion centre for the Vallée de Joux (lakes surrounded by gentle pastures, backed by steeper hills used for skiing) and for fishing in the little Orbe river or walks to the cascading source of the Orbe. Here you may visit **Les Grottes de Vallorbe**, and tour the world of stalactites, stalagmites, underground torrents and lagoons.

INDEX

This index is in three separate parts. The first part (below) refers to all the general information in the book. Each country has its own index which refers to the gazetteer. In the last two indexes all the main entries are printed in heavy type. Map references are also printed in heavy type. The map page number precedes the grid reference.

Accommodation	20–2	
Air services	14, 16	
internal	16	
Alps	8	
Architecture	12	
Arts	10–12	
Babenbergs	6, 39, 49, 59	
Bands	27	
Bank opening hours	14	
Beer	23	
Beethoven	11, 40, 45	
Bicycles	17	
Bobsledding	26	
Brahms	45	
Breakdowns	20	
Breathalyzer	20	
Bruckner	11, 38, 65	
Bus services	16	
internal	17	
Cable car	18	
Cable chair	18	
Camping	22	
Canoeing	24	
Car *see* Motoring		
Car ferries and routes	16	
Car rental	18	
Caravans (trailers)	20	
Casinos	27	
Caving	24	
Chemist	28	
Churches	28	
Cigarettes	28	
Climbing	24	
Coach *see* Bus services		
Coffee	23	
Composers	10–11	
Consulates, addresses	32	
Credit cards	14	
Currency	14	
Customs	14	
duty-free allowances	13	
Drinks	23	
Eating places	22–3	
Electricity	28	
Embassies, addresses	32	
Emergency medical treatment		28
Emergency police telephone	29	

Eurocheque card	14	
Exchange offices	14	
Farmhouse accommodation	21	
Ferdinand I	7	
Festivals	27–8	
First-aid treatment	13	
Fishing	24	
Flights *see* Air services		
Folklore	27–8	
Food and drink	22–4	
Franz II (I of Austria)	7, 39	
Franz-Josef	7–8, 35, 58, 64	
Frederick II	6	
Frederick III	7	
Funicular	17–18	
Gasoline	20	
Gasthof	21	
Gluck	11	
Golf	25	
Gondola	18	
Government	9	
Gruber	11, 64	
Habsburgs	6–8, 34, 44, 45, 59, 92, 110	
Haydn	11, 48	
Health care	28	
Heurigen	23	
Hiking	26	
History	6–8	
Hitchhiking	17	
Hobby holidays	25	
Hohenstaufen dynasty	6	
Hotels	20–1	
grades	21	
price reductions	21	
Industry	10	
Inoculations	13	
Insurance:		
medical	13	
property	13	
Johann, Archduke	52, 56	
Klee, Paul	12, 44	
Language	32–3	
Lehár	11	
Leopold I	7	

Liszt	11, 48	
Literature	12	
Lost property	29	
Mahler	11	
Maria Theresia	7, 45, 73, 74	
Maximilian I	7	
Medical insurance	13	
Medical treatment	28	
Motels	21	
Motoring	18–20	
alcohol	20	
breakdowns	20	
documents	18	
lights	18	
parking	20	
petrol	20	
road conditions	19–20	
rules of the road	18	
speed limits	19	
tyre pressure	30	
use of horn	20	
Motoring organizations	20	
addresses	32	
Mountain roads and passes		19–20
Mountain transport	17–18	
warnings	18	
Mozart	11, 40, 45, 65–6, 67	
Museum opening times	29	
Music:		
centres for classical	26–7	
composers	10–11	
jazz	27	
Newspapers	29	
Nightclubs	27	
Parking	20	
Passports	13	
Petrol	20	
Photography	29	
Police	29	
Population	8	
Postal services	29	
Public holidays	29–30	
Rabies	28–9	
Radio	30	
Rail services	16	
internal	16–17	
Regions	9–10	

Index 125

Religion	9	
Restaurants	22–3	
Riding	25	
Ring of the Nibelung	50	
Road conditions	19–20	
Rudolf of Habsburg	6	
Rudolf II	7	
Sailing (dinghy)	25	
Sailings	16	
Danube	17	
lake trips and services	17	
Schubert	11, 38, 40, 45, 51	
Self-catering accommodation	21	
Ships *see* Sailings		
Shop opening times	29	
Shopping	30	
Skating	25	
Ski lift	18	
Skibob	26	
Skiing	25–6	
après-ski	26	
cross-country	26	

downhill	25
some guide prices	25
summer	26
Spas	22
Speed limits	19
Spirits	23
Sports	24–6
Strausses	11, 38, 40, 45
Swimming	26
Swiss army	9
Taxis	18
Telegrams	31
Telephones	30–1
some useful codes	31
Tennis	26
Theatre	27
Time	30
Tipping	30
Tobacco	28
Tobogganing	26
Toilets	30
Tourist organization addresses, abroad:	

Austrian national	31
Swiss national	31
Tourist organization addresses, internal:	
Austrian provincial	32
Swiss regional	31–2
Tourist organizations	9
Trailers	20
Trains *see* Rail services	
Travel	14, 16
internal	16–18
price reductions	17
Travellers' checks	14
Visas	12–13
Walking	26
Water	30
Wines	23
Work permit	12
Writers	12
Yodelling	27
Youth hostels	21–2

AUSTRIA

Admont	52–3, 54–5 **E13**
Alberschwende	71
Alps	48, 52, 70
Andau	50
Andelsbuch	71
Annaberg	59
Axalm	74
Bad Aussee	53, 54–5 **E10**
Bad Hofgastein	61
Bad Ischl	62–3, **K8**, 64
Bad Tatzmannsdorf	46–7 **N10**, 51
Bad Vöslau	51
Baden	46–7 **H11**, 48
Badgastein	61, 62–3 **P5**
Bartholomaberg	71
Bernstein	51
Berwang	77
Bezau	71, 72–3 **D5**
Bludenz	71, 72–3 **F4**
Bodensee	71–2
Brand	71
Brandner valley	71
Braunau	65
Bregenz	71–2, 72–3 **C4**
Bregenzerwald	70, 71
Bruck an der Mur	53, 54–5 **G18**
Burgenland	46
Carinthia (Kärnten)	52
Constance, Lake	71–2

Danube (Donau)	38, 45, 46, 48, 61, 62
Danube Canal	35, 38
Danube, Old	45
Döllach	58
Donawitz	53, 54–5 **G17**
Dornbirn	72, 72–3 **C4**
Dürnstein	49
Eastern Austria	46–51, 46–7
festivals, etc	48
food (Burgenland)	48
Egg	71
Ehrwald	77
Eisenerz	71
Eisenstadt	46–7 **J12**, 48
Erl	76
Erzberg (Mt.)	53
Felbertauern tunnel and road	52, 58
Feldkirch	72, 72–3 **E3**
Finkenberg	77
Fontanella	71
Forchtenstein	46–7 **K11**, 48
Freistadt	61–2, 62–3 **D13**
Fulpmes	78
Gaming	59
Gänsendorf	46–7 **F13**, 48
Gargellen	71

Gasberg	68
Gaschurn-Partenen	71
Gesäuse gorge	53
Gloggnitz	51
Gmund	58
Gmunden	62, 62–3 **I9**
Gortipohl	71
Göstling	59
Graz	53, 54–5 **J19**, 56
Great Walser valley	71
Grein	62, 62–3 **F15**
Gross Venediger (Mt.)	52, 60
Grossdorf	71
Grossglockner (Mt.)	52, 56, 58
Grundl, Lake	53
Gumpoldskirchen	51
Gurk	54–5 **K13**, 56
Güssing	51
Haag	65
Hall in Tirol	72–3, 72–3 **E14**
Hallein	62, 62–3 **K5**, 64
Hallstatt	62–3 **L8**, 64
Hart	77
Heiligenblut	54–5 **J4**, 56, 58
Hellbrunn Palace	68
Hinterbruhl	51
Hochgurgl	78
Hochosterwitz	54–5 **L13**, 58
Hochsölden	78
Hopfgarten	76

126 Index

Igls 74
Ill valley 71
Illmitz 50
Imst 72–3 E9, 73
Inn valleys 70, 76, 77
Innsbruck 72–3 E13, 73–4
Innviertel 61, 65

Johnsbach 53

Kaisergebirge 76
Kaprun 69
Kärnten (Carinthia) 52
Katerloch 56
Kirchberg 76
Kirchdorf 76
Kitzbühel 72–3 D19, 76
Klagenfurt 54–5 M13, 58–9
Klosterneuburg 46–7 F11, 48–9
Krems 46–7 E7, 49
Krimml waterfalls 65
Kufstein 72–3 B18, 76

Landeck 72–3 F9, 76
Längenfeld 78
Lech 72–3 E6, 76
Lechtal Alps 70
Leoben 53
Leopoldskron Palace 68
Lermoos 77
Liechtenstein (**Vaduz**) 70–1, 72–3 F3, 79
Lienz 54–5 L4, 59
Lilienfeld 59
Linz 62–3 F12, 65
Lofer 62–3 L3, 65
Lungau 69
Lunz 59

Marchfeld 48
Maria Plain 68
Maria Wörth 54–5 N12, 59
Mariazell 54–5 D18, 59
Matrei am Brenner 60
Matrei in Osttirol 54–5 J3, 59–60
Mattsee 65
Mauterndorf 69
Mauthausen 65
Mayerling 51
Mayrhofen 72–3 F16, 76–7
Melk 46–7 E7, 49–50
Mellau 71
Mieders 78
Mils 73
Milstatt 54–5 L9, 60
Mittersill 62–3 N2, 65
Mödling 51
Montafon 70, 71
Moosham Castle 69
Mörbisch 50
Mühlviertel 61
Mutters 74

Natters 74
Neusiedl, Lake 50
Neusiedl am See 46–7 J14, 50
Neustift 78
Niederösterreich 46
Northern Austria 61–9, 62–3

Obergurgl 78
Oberndorf (Salzburg) 68
Oberndorf (Tirol) 76
Oberösterreich 61
Obertauern 69
Osttirol 52
Ötscher massif 48, 59
Ötz 78
Ötz valley 78

Pamhagen 50
Peggau 56
Perchtoldsdorf 51
Petronell 46–7 H14, 50
Piber 56
Pöchlarn 50
Podersdorf 50
Präbichl 53
Prägraten 60
Puchberg 51

Radstadt 62–3 N8, 65
Ragga Gorge 60
Rattenberg 72–3 D16, 77
Raxalpe 48, 51
Reichenau 51
Retz 51
Ried 77
Ried im Innkreis 62–3 G7, 65
Rohr-im-Gebirge 51
Rust 46–7 J13, 50

St Anton 71
St Anton am Arlberg 72–3 F7, 77
St Florian 65
St Gilgen 62–3 J7, 69
St Jakob im Defereggental 54–5 K2, 60
St Johann in Tirol 72–3 C19, 77
St Margarethen 50
St Oswald 62
St Ulrich 76
St Veit an der Glan 54–5 L13, 60
St Wolfgang 62–3 J8, 69
Salzburg 62–3 J5, 65–8
Salzburg province 61
Salzkammergut 61
Sautens 78
Schallaburg castle 50
Schloss Eggenberg 56
Schmittenhöhe 69
Schneeberg 48, 51

Schoppernau 71
Schröcken 71
Schruns 71
Seefeld 72–3 D12, 77
Semmering 46–7 L9, 50–1
Silvretta range 71
Sölden 72–3 G11, 78
Sonntag 71
Southern Austria 52–60, 54–5
Spittal (an der Drau) 54–5 L8, 60
Sportgastein 61
Steiermark (Styria) 52
Steyr 62–3 H13, 69
Stubai Tal 72–3 F13, 78
Stubing 56
Stumm 77

Tamsweg 62–3 P10, 69
Tauern range 52, 58
Tirol 70
Tirol, East 52
Trausdorf 50
Tschagguns 71
Tulfes 74
Tulln 46–7 F9, 51

Uderns 77
Umhausen 78
Untersberg 68

Vent 78
Vienna (Wien) 34–45
  greater city 44–5
  greater city map 42–3
  history 34–5
  inner city 39–41, 44
  inner city map 36–7
  The Ring 35, 38
Vienna Woods see Wienerwald
Villach 54–5 M10, 60
Virgen 60
Völkermarkt 54–5 M14, 60
Vorarlberg province 70
Vordernberg 53

Wachau 46, 48, 49
Walchsee 76
Waldviertel 46–7 C5, 48, 51
Weiden 50
Weinviertel 48
Wels 62–3 G11, 65
Werfen 64
Western Austria 70–8, 72–3
Wien see Vienna
Wienerwald 46, 46–7 H9, 51
Wienerwald village 51

Zell am See 62–3 N4, 69
Zell am Ziller 77
Ziller valley 70, 76–7
Zistersdorf 48
Zwettl 51

Index

SWITZERLAND

Aarau 103, 104–5 **J9**	Delemont 114–15 C13, 116	Jungfraujoch 97
Aare Gorge 101	**Les Diablerets** 114–15 **P11**,	Jura mountains 103, 116, 123
Adelboden 93	116	
Airolo 85 **A4**, 87	Disentis 80–1 **H2**, 83	**Kandersteg** 94–5 **M7**, 98
Alpnach Brunnen 101	Doubs, R. 116, 123	Kleine Scheidegg 97
Alps 82, 85–6, 88, 92		Klosters 83
Altstätten 103, 104–5 **J20**	Ebnat-Kappel 113	**Kreuzlingen** 104–5 **F20**,
Andermatt 92–3, 94–5 **L15**	**Einsiedeln** 93, 94–5 **F16**	110
Appenzell 103, 104–5 **J18**,	Elm 93	Kussnacht 101
106	Engadine (En valley) 82	
Arolla 88–9 **H7**, 89	**Engelberg** 93, 94–5 **I13**	**Langnau** 94–5 **H8**, 98
Arosa 80–1 **G9**, 83	Estaveyer le Lac	**Lausanne** 114–15 **M7**, 122
Ascona 85 **F5**, 87	114–15 **I9**, 116	Lauterbrunnen 97
Augusta Rurica 106	Evolene 88–9 **G7**, 90	Léman, Lac 114, 116
Les Avants 123		Lenk 102
Avenches 123	Faido 85 **A6**, 87	Lenzerheide 83
	Fiesch 88–9 **C12**, 90	Leukerbad 88–9 **C8**, 90
Bad Ragaz 104–5 **N20**, 110	**Flims** 80–1 **G5**, 83	**Leysin** 114–15 **O10**, 122
Baden 104–5 **H11**, 106	Fribourg (Freiburg)	Linthal 93
Basel 104–5 **G5**, 106	114–15 **J11**, 116–17	Locarno 85 **E6**, 87
Beatenberg 102	**Frutigen** 93, 94–5 **L7**	**Lugano** 85 **G7**, 87
Beatenbucht 102		**Luzern** 94–5 **F12**, 99–101
Bellinzona 85 **E8**, 87	Gambarogno 85 **F5**, 87	Luzern area 92
Bern 104–5 **O4**, 107	Geneva, Lake 114, 116	
town plan 108–9	Genève (Geneva)	**Martigny** 88–9 **G3**, 90
Berner Oberland 92	114–15 **Q2**, 118–19	**Meiringen** 94–5 **K11**, 102
Bernina Pass 83	town plan 120–1	Merlingen 102
Bettmeralp 90	Gersau 101	Mittelland 103, 116
Biel (Bienne) 104–5 **L2**,	Giessbach falls 93	Montana see Crans/Montana
107, 110	Giornico 87	**Montreux** 114–15 **N9**,
Blonay 122	Glarus 93, 94–5 **G19**	122–3
Bodensee 110	**Gletsch** 88–9 **C12**, 90	**Morges** 114–15 **M6**, 123
Bougy-Villars 123	Glion 123	**Muotathal** 94–5 **G16**, 101
Braunwald 93	Göschenen 92–3	Mürren 97
Brienz 93, 94–5 **J10**	**Graubünden (Grisons)**	Murten (Morat)
Brienzer Rothorn 93	80, 80–1, 82–4	114–15 **M11**, 123
Brig 88–9 **D11**, 89	festivals 83	
Brione 85–6	language 82	National Park 82, 84
Brissago 85 **F5**, 87	leagues 82–3	**Neuchâtel** 114–15 **H9**, 123
Brugg 104–5 **H10**, 110	Great St Bernard Pass 90	**La Neuveville** 114–15 **G11**,
Buchs 113	Griments 91	123
Burgenstock (Mt.) 100	Grindelwald 96–7	**Le Noirmont** 114–15 **E10**,
	Gruyères 114–15 **M11**,	123
Caux 123	119, 122	**Northern Switzerland**
Central (Heartland)	Gryon 114–15 **P10**, 119	103–13, 104–5
Switzerland 92–102,	Gstaad 102	
94–5	Guarda 84	**Olten** 104–5 **I7**, 110
Champery 88–9 **F2**, 89		Onsernone valley 86
Champex 88–9 **H4**, 90	Les Haudières 90	
Château d'Oex 114–15 **N11**,	Hergiswil 101	Payern 116
116	Herisau 106	Pilatus (Mt.) 100
La Chaux de Fonds	Hilterfingen 102	Planplatten 101
114–15 **F9**, 116	Holloch grotto 102	Les Pleiades 122–3
Chexbres 114–15 **N8**, 116		Pontresina 84
Chillon, Château de 122	Ilanz 83	**Poschiavo** 80–1 **M12**, 83
Chironico 87	Immensee 101	
Chur 80–1 **G8**, 83	Inn river see Engadine	Rappersvil 113
Constance, Lake 110	**Interlaken** 93, 94–5 **K8**,	Rhine 82, 84, 111, 112
Crans/Montana 88–9 **D7**, 90	96	Rhône 88, 90
		Riederalp 90
Davos 80–1 **G10**, 83	Jungfrau 94–5 **K8**, 96–7	Rigi (Mt.) 100

Rochers de Naye 123
Rolle 114–15 **N4**, 123
Ronco All'Acqua 87
Ronco Ascona 85 **F5**, 87
Rorschach 104–5 **H21**, 110
Rosenlaui 101

Saanen 102
Saanenmöser 102
Saas Fee 88–9 **G11**, 90
St Gallen 104–5 **H20**, 110
St Gotthard pass/tunnels
85, 87, 92
St Maurice 88–9 **E3**, 90
St Moritz 80–1 **K10**, 83–4
St Stephen 102
St Ursanne 114–15 **C11**, 123
Ste Croix 114–15 **J6**, 123
Samedan 84
San Bernardino pass 84, 85, 87
Sarnen 94–5 **H12**, 101
Schaffhausen 104–5 **E14**, 110–11
Schilthorn 97
Schleitheim 111

Schönried 102
Schwyz 94–5 **G15**, 101–2
Scuol 80–1 **G14**, 84
Seelisberg 101
Sempach 94–5 **E11**, 102
Sierre 88–9 **D7**, 90
Silvaplana 84
Simmental 94–5 **L5**, 102
Simplon pass/tunnels 89
Sion 88–9 **E6**, 90
Sisikon 101
Solothurn 104–5 **K4**, 111
Southern Switzerland see
 Graubünden, Ticino,
 Valais
Spiez 102
Splügen 80–1 **J6**, 84
Stansstad 101
Stein am Rhein 104–5 **F16**, 112
Stoos 102
Sursee 102

Tarasp 84
La Tène 123
Thun 94–5 **J6**, 102
Thusis 80–1 **H7**, 84

Ticino 80, 85–7, 85
 food and drink 86–7
Toggenburg 104–5 **J17**, 113
Trogen 106

Val de Travers 123
Valais 80, 88–91, 88–9
 food and drink 89
Vallorbe 114–15 **K5**, 123
Verbier 88–9 **G5**, 91
Vitznau 101
Vulpera 84

Walensee 110
Weggis 101
Wengen 97
Western Switzerland
114–23, 114–15
Wildhausen 113
Winterthur 104–5 **H15**, 113

Zermatt 88–9 **H9**, 91
Zernez 80–1 **I12**, 84
Zinal 88–9 **F8**, 91
Zug 94–5 **E14**, 102
Zuoz 80–1 **F11**, 84
Zürich 104–5 **I13**, 113
Zweisimmen 102